SpringerBriefs in Molecular Science

SpringerBriefs in Molecular Science present concise summaries of cutting-edge research and practical applications across a wide spectrum of fields centered around chemistry. Featuring compact volumes of 50 to 125 pages, the series covers a range of content from professional to academic. Typical topics might include:

- A timely report of state-of-the-art analytical techniques
- A bridge between new research results, as published in journal articles, and a contextual literature review
- A snapshot of a hot or emerging topic
- An in-depth case study
- A presentation of core concepts that students must understand in order to make independent contributions

Briefs allow authors to present their ideas and readers to absorb them with minimal time investment. Briefs will be published as part of Springer's eBook collection, with millions of users worldwide. In addition, Briefs will be available for individual print and electronic purchase. Briefs are characterized by fast, global electronic dissemination, standard publishing contracts, easy-to-use manuscript preparation and formatting guidelines, and expedited production schedules. Both solicited and unsolicited manuscripts are considered for publication in this series.

Pratima Bajpai

Chimeric Enzymes Showing Promise in Pulp and Paper Sector

Pratima Bajpai
Pulp and Paper Consultant
Kanpur, Uttar Pradesh, India

ISSN 2191-5407   ISSN 2191-5415 (electronic)
SpringerBriefs in Molecular Science
ISBN 978-3-032-18002-5   ISBN 978-3-032-18003-2 (eBook)
https://doi.org/10.1007/978-3-032-18003-2

This Springer imprint is published by the registered company Springer Nature Switzerland AG
The registered company address is: Gewerbestrasse 11, 6330 Cham, Switzerland

Preface

This SpringerBrief offers clear perspectives on the targeted development of chimeric enzymes using enzyme engineering strategies for their efficient application in the pulp and paper sector. The pulp and paper sector is crucial for fostering a country's social and economic development by providing necessary goods. Traditional pulp and paper processing typically requires a large amount of energy and employs chemical substances that can generate harmful intermediates. The application of enzymes in the pulp and paper sector has greatly decreased the chemical use and energy consumption throughout the processing stages. Different enzyme cocktails are utilised to carry out several tasks in a single process, but because of significant variations in operating conditions, they frequently do not function in concert. The need for designed chimeric enzymes is highlighted by the lack of synergy across various operating settings. Additionally, enzymes can function effectively even under less-than-ideal circumstances thanks to enzyme engineering approaches. To increase their catalytic potential and stability while in use, enzymes have undergone modifications. The development of chimeric or multifunctional enzymes can allow for simultaneous activity in a variety of operating conditions. In the pulp and paper business, chimeric enzymes help many enzymes work together more effectively. This book will be useful for Pulp and Paper technologists, Biotechnologists, Biochemist, Applied Chemists and Chemical engineers.

Kanpur, India

Pratima Bajpai

Acknowledgments I wish to express my sincere gratitude to the many individuals and organizations who generously contributed valuable information to this work. I am deeply appreciative of the various publishers who kindly granted permission to reproduce or adapt their content. My heartfelt thanks are extended to Elsevier, Springer, and the numerous open-access journals and publications, whose support and resources have been indispensable in the completion of this study.

Contents

1 General Background 1
   1.1 Market Trends in Pulp and Paper Industry 1
   1.2 Environmental and Energy Challenges in Conventional
       Production 2
   1.3 Enzyme Innovations and Chimeric Solutions 3
   References 8

2 Pulp and Paper Making Process 11
   2.1 Global Status of the Pulp and Paper Industry 11
   2.2 Manufacturing Process 14
   References 18

**3 Utilization of Enzymes and Enzyme Blends Within the Pulp
and Paper Sector** 19
   3.1 Enzymes and Supporting Proteins in Pulp and Paper
       Processing 21
   3.2 Enzyme Cocktails in Pulp and Paper Processing 26
   3.3 Factors Influencing Enzyme Cocktail Performance 36
   3.4 Advancements in Biotechnological Practices in the Pulp
       and Paper Field 40
   3.5 Future Considerations 41
   References 41

4 Chimeric Enzymes 49
   4.1 Chimeric Enzymes: A New Frontier in Protein Engineering 49
      4.1.1 Chimeric Enzymes Found in Nature 50
      4.1.2 Artificially Created or Recombinant Chimeras 51
   4.2 Design of Chimeric Enzymes to Enhance Stability, Activity,
       and Specificity 58
      4.2.1 Improvement in Stability 58
      4.2.2 Improvement in Activity 60
      4.2.3 Merging and Altering the Specificity 61

4.3   Advantages of Utilizing Chimeric Enzymes for Catalytic
       Processes   64
       4.3.1   Introduction   64
       4.3.2   Definition and Concept of Chimeric Enzymes   64
       4.3.3   Applications in Medicine and Biotechnology   64
       4.3.4   Industrial and Biotechnological Relevance   65
       4.3.5   Design Considerations and Challenges   67
4.4   Designing Strategies for Chimeric Enzymes   68
       4.4.1   Domain Swapping and Fusion in Chimeric Enzyme
                Design   69
       4.4.2   Active Site Preservation in Chimeric Enzyme Design   71
       4.4.3   Scaffold Selection in Chimeric Enzyme Design   75
       4.4.4   Surface Residue Optimization in Chimeric Enzyme
                Design   76
       4.4.5   Computational Modeling and Simulation in Chimeric
                Enzyme Design   78
       4.4.6   Linker Engineering in Chimeric Enzymes   80
4.5   Engineering Enzymes to Improve Chimeric Enzyme
       Functionality   81
       4.5.1   Why Chimeric Enzymes?   81
       4.5.2   Principles of Chimeric Enzyme Design   82
       4.5.3   Strategies to Improve Functionality   83
       4.5.4   Representative Case Studies   84
       4.5.5   Experimental Workflows and Practical
                Considerations   85
       4.5.6   Challenges and Limitations   86
       4.5.7   Future Directions   86
4.6   Potential Applications of Chimeric Enzymes in Pulp
       and Paper Industry   87
       4.6.1   Limitations of Conventional Chemical Treatments   87
       4.6.2   Emergence of Enzymatic Alternatives   88
       4.6.3   Chimeric Enzymes as Next-Generation Biocatalysts   88
       4.6.4   Technical Design Considerations   89
       4.6.5   Expression Systems and Production Platforms   89
       4.6.6   Industrial Benefits of Chimeric Enzymes   90
       4.6.7   Enzymatic Deconstruction of Lignocellulosic
                Biomass   90
       4.6.8   Case Studies of Chimeric Enzyme Applications   91
       4.6.9   Advances in Thermostability and Catalytic Efficiency   92
       4.6.10   Regulatory and Transcriptional Chimeras   93
       4.6.11   Industrial Bio-Bleaching Applications   93
       4.6.12   Challenges and Limitations   101

4.7 Forward-Looking Perspectives on Chimeric Enzymes
for Sustainable Paper Production 104
   4.7.1 Market Outlook for Enzyme Technologies 104
   4.7.2 Future Perspectives 105
References 109

Index 121

List of Figures

Fig. 2.1   Process flow in Pulp and Paper manufacturing mills. Haile et al. (2021). Licensed under a Creative Commons Attribution 4.0 International License 16

Fig. 3.1   Representation of enzyme applications within P&P processes. The figure's notes highlight the moments in the process when enzymes can be added. Yang et al. (2023). Reproduced with permission 20

Fig. 4.1   A schematic illustration of the three techniques for fusing domains or proteins: posttranslational conjugation (**c**), domain insertion (**b**), and end-to-end gene fusion (**a**). Jaduan et al. (2020). Reproduced with permission 52

Fig. 4.2   A description of the various forms of domain insertions in proteins: **a** single insertion, **b** nested insertion, **c** two-domain insertion, and **d** three-domain insertion. N and C indicate the N- and C-terminal ends of the parent domain or protein, respectively. Jaduan et al. (2020). Reproduced with permission 53

Fig. 4.3   Natural multidomain chimeric enzymes. Examples of the seven different kinds of multidomain enzymes found in nature. García-Paz et al. (2024). Licensed under a Creative Commons Attribution 4.0 International License 57

Fig. 4.4   Two fundamental categories of fusion, or chimeric, proteins. The first category comprises two proteins or protein subunits that are fused end-to-end, typically connected by a linker. The second category involves the interspersion of amino acids from both donor proteins within the resulting fusion protein product. Strohl (2017). Reproduced with permission 58

List of Tables

Table 1.1   Comparative benefits of chimeric versus. Natural enzymes in pulp and paper 7

Table 1.2   Applications of chimeric enzymes in pulp and paper processes 7

Table 1.3   Strategic advantages of chimeric enzyme integration 8

Table 1.4   Challenges and enablers for chimeric enzyme adoption 8

Table 3.1   Enzymes used in pulp and paper processing 22

Table 3.2   Enzyme applications in biobleaching in recent years 29

Table 3.3   Use of chemical additives and enzymes in the papermaking process 34

Table 3.4   Readiness levels of enzyme technologies in pulp and paper 37

Table 3.5   Global enzyme manufacturers serving the pulp and paper market 38

Table 3.6   Industrial Deployment Status of Enzymes in Papermaking and Leading Market Players 39

Table 4.1   Chimeric synthetic enzymes demonstrating superior physicochemical properties relative to native enzymes 55

Table 4.2   Design considerations for chimeric enzymes 62

Table 4.3   Chimeric enzyme constructs and functional enhancements 63

Table 4.4   Design factors for chimeric enzymes 81

Table 4.5   Chimeric enzymes exhibiting possible uses in the pulp and paper sector 96

Table 4.6   Technical challenges and limitations of chimeric enzymes 104

Table 4.7   Future insights on chimeric enzymes 108

Chapter 1
General Background

Abstract The pulp and paper industry (PPI) is transitioning toward sustainable, energy-efficient manufacturing. Conventional chemical-intensive processes in pulping, bleaching, and recycling raise environmental and economic concerns, driving adoption of biotechnological alternatives. Chimeric enzymes—created by fusing catalytic and binding domains—offer superior stability, specificity, and tolerance compared to natural enzymes. In PPI, they enable biobleaching, deinking, pitch control, and fiber modification, reducing chemical use, improving pulp quality, and lowering energy demand. Their role extends to biorefineries, facilitating efficient lignocellulosic biomass conversion. Advances in protein engineering and synthetic biology position chimeric enzymes as key drivers of green, circular, and economically viable production.

Keywords Pulp and paper industry · Chimeric enzymes · Biobleaching · Deinking · Protein engineering · Lignocellulosic biomass · Enzyme stability · Sustainable processing · Pitch control · Industrial biotechnology

1.1 Market Trends in Pulp and Paper Industry

The pulp and paper industry (PPI) represents a cornerstone commodity sector that is intricately linked to global economic progress, industrial development, and social well-being. It not only supplies essential products such as paper, packaging materials, and tissue but also underpins communication, education, and trade across the world. Ranking among the largest industrial domains globally, the PPI has historically been dominated by production hubs in North America, Northern Europe, and East Asia, regions that established early leadership through technological innovation, abundant forest resources, and well-developed infrastructure. In addition, Latin America and Australia have emerged as important contributors, leveraging their resource bases and expanding export capacity to meet international demand.

P. Bajpai, *Chimeric Enzymes Showing Promise in Pulp and Paper Sector*,
SpringerBriefs in Molecular Science, https://doi.org/10.1007/978-3-032-18003-2_1

As mature markets in Europe and North America approach saturation, growth trajectories in these regions are expected to plateau, with limited scope for significant expansion beyond incremental efficiency gains and product diversification. By contrast, the Asia–Pacific region—particularly India and China—has become the focal point of future industry growth. Rising populations, rapid urbanization, and increasing literacy rates are driving demand for paper products ranging from packaging to printing and specialty grades. This demographic and economic momentum is catalyzing substantial expansion in the global paper industry, positioning Asia as the epicenter of production and consumption in the coming decades.

At present, Asia accounts for approximately 40% of global paper and paperboard output, underscoring its dominant role in shaping industry dynamics. The European Union and North America each contribute around 25%, reflecting their continued importance despite slower growth. Together, these regions highlight the evolving balance of global production, where traditional leaders maintain strong industrial bases while emerging economies redefine the future landscape. The shifting geography of pulp and paper manufacturing illustrates not only economic transitions but also the broader challenges and opportunities associated with sustainability, resource management, and technological innovation in a sector that remains vital to global development.

In recent years, profitability across the global paper sector has faced mounting pressure due to overcapacity, declining prices, and escalating production costs. Trade volumes remain robust, with the most significant flows of paper and paperboard occurring between China and Europe—particularly Northern Europe—and between Canada and the United States. South Korea also stands out as a prominent exporter within the Asian market.

1.2   Environmental and Energy Challenges in Conventional Production

Despite its economic importance, the PPI is inherently resource- and energy-intensive, demanding substantial inputs of water, electricity, and chemicals. Conventional production methods generate approximately 950 kg of CO_2-equivalent greenhouse gas emissions per ton of paper, accounting for 5.7% of global final industrial energy consumption and 9% of greenhouse gas emissions from the manufacturing sector (Sun et al. 2018; Yakubu et al. 2019). The industry is frequently cited as a contributor to environmental degradation, including carbon emissions, water contamination, solid waste generation, and climate change (Söderholm et al. 2019).

In light of global carbon neutrality targets and the increasing adoption of renewable energy and cleaner production technologies, the transformation of the PPI toward green, low-carbon, and sustainable models is both urgent and essential. One promising avenue for such transformation is the replacement of traditional chemical and mechanical processes with enzyme-based catalysis. Although enzyme applications in the PPI were first reported in 1984 (Paice and Jurasek 1984), early formulations were hindered by high costs, limited recyclability, and poor stability. However, advancements in enzyme-centric industrial biotechnology over the past few decades have markedly enhanced enzyme performance and cost-effectiveness. Today, a wide array of enzyme treatments are actively employed across various pulp and paper (P&P) processes.

1.3 Enzyme Innovations and Chimeric Solutions

Emerging sub-sectors within the pulp and paper industry (PPI), notably dissolving pulp and nanocellulose, present significant opportunities for enzyme-based innovations and are anticipated to play a central role in advancing sustainability and technological progress (Kumar and Christopher 2017; Ribeiro et al. 2019). The integration of enzymes into PPI operations has markedly contributed to the industry's environmental transformation—enhancing pulp yield and improving paper properties while simultaneously reducing energy consumption and minimizing ecological impact (Bajpai 2011, 2018; Chaurasia and Bhardwaj 2019; Sharma et al. 2020; Wei et al. 2021; Yang et al. 2023; Henríquez-Gallegos et al. 2024; Lee et al. 2023; Onwosi et al. 2024). The commercial landscape for PPI-related enzymes is expanding rapidly, with major global suppliers such as Novozymes, DuPont, Buckman, DSM, and AB Enzymes leading the market. Widely adopted enzyme formulations include Fiber-Care® and Paperzyme® (Novozymes), Accellerase® (DuPont), Fibrolytic enzymes (AB Enzymes), Paperzyme (Amano Enzyme), and Maximyze® (Buckman) for refining; BioDeink® N (Novozymes) and Enzynk® (Enzymatic Deinking Technologies) for deinking; PrimaGreen® (DSM) for bleaching; and StickAway®, Resinase® A 2X (Novozymes), Cellulase (DuPont), and Optimyze® (Buckman) for sticky control in recycled paper processing.

Chimeric enzymes have emerged as a transformative innovation in the pulp and paper industry, offering not only enhanced catalytic efficiency but also significant strides toward environmental sustainability and process optimization (García-Paz et al. 2024; Arsalan et al. 2025; Monica et al. 2024; Shahid et al. 2023). These engineered proteins are constructed by fusing functional domains from distinct parental enzymes, resulting in multifunctional biocatalysts capable of performing complex tasks that natural enzymes often cannot. By integrating catalytic and binding modules with complementary properties, chimeric enzymes overcome several inherent limitations of native enzymes—such as low thermal stability, narrow substrate specificity, and limited tolerance to industrial stressors. In pulp and paper manufacturing, where operations frequently involve extreme pH conditions, elevated temperatures, and the presence of inhibitory compounds like lignin derivatives and metal ions, the robustness and adaptability of chimeric enzymes make them particularly valuable. Their ability to retain activity under such harsh conditions enables more efficient hydrolysis of lignocellulosic materials, which are the primary constituents of wood pulp.

This enhanced breakdown not only improves process throughput but also reduces energy consumption and facilitates more consistent product quality.

In recent years, there has been a growing demand for enzymes with enhanced activity, selectivity, and stability across diverse industrial sectors. Through strategic engineering approaches, enzymes can be tailored for specific applications including food processing, animal nutrition, pharmaceuticals, therapeutics, bioremediation, biofuels, and detergents (da Silva et al. 2022). One such innovation involves the synthesis of multifunctional proteins by integrating desirable traits from multiple proteins into a single peptide construct (Ma et al. 2020). Chimeric enzymes—also referred to as hybrid enzymes—are formed by merging two or more independent genes, resulting in a translational product that combines distinct functional domains. These domains may originate from different sources, serve varied biological roles, or reside in separate cellular compartments.

Chimeric enzyme systems offer a robust solution to common limitations associated with enzyme cocktails, particularly issues of inactivation and incompatibility. By enhancing the structural stability of constituent enzymes, chimeric constructs enable more reliable and synergistic performance. Several chimeric enzymes with lignocellulolytic activity have demonstrated superior efficiency in degrading lignocellulosic biomass compared to conventional multi-enzyme treatments (Shahid et al. 2023). To further improve industrial applicability, researchers have employed advanced techniques such as directed evolution, semi-rational design, immobilization engineering, and whispering-gallery mode sensors to optimize catalytic performance and durability.

In the context of the pulp and paper industry, chimeric enzymes offer distinct advantages due to the sequential nature of enzymatic treatments required during processing. Traditional enzyme cocktails often fall short in such scenarios due to incompatibility among individual components. Chimeric enzymes, by contrast, provide integrated functionality and enhanced resilience, making them particularly well-suited for complex, multi-step industrial workflows.

Traditionally, the pulp and paper production process relies heavily on chemical treatments for pulping, bleaching, and fiber modification. These treatments—especially those involving chlorine-based compounds and strong alkalis—contribute significantly to environmental pollution, generate toxic effluents, and pose health risks to workers and surrounding communities. The discharge of chlorinated organics and other persistent pollutants into water bodies has long been a concern for regulatory agencies and environmental advocates. In this context, enzymatic alternatives—particularly those involving chimeric enzymes—offer a more sustainable, targeted, and biologically compatible approach by reducing the dependency on aggressive chemical reagents. For instance, chimeric xylanases have demonstrated superior performance in enhancing the bleaching process by selectively degrading hemicellulose and facilitating lignin removal. This enzymatic action improves pulp brightness, reduces effluent toxicity, and minimizes the environmental footprint of the bleaching stage. Additionally, the use of such enzymes can improve fiber integrity, reduce the need for secondary chemical treatments, and contribute to better mechanical properties of the final paper product, such as tensile strength, opacity, and surface smoothness.

Moreover, the modular nature of chimeric enzymes allows for fine-tuning of their activity profiles to match specific process requirements. For example, enzymes can be engineered to target particular hemicellulose linkages or to function optimally at the pH and temperature conditions prevalent in industrial digesters. This level of customization is particularly advantageous in mills processing diverse feedstocks, such as hardwoods, softwoods, or agricultural residues, each of which presents unique enzymatic challenges. By integrating chimeric enzymes into existing workflows, manufacturers can achieve greater consistency in pulp quality, reduce variability in bleaching outcomes, and lower operational costs associated with chemical procurement and effluent treatment. As the industry continues to prioritize sustainability and regulatory compliance, the adoption of chimeric enzyme technologies represents a strategic shift toward greener, more resilient manufacturing practices.

Beyond bleaching, chimeric enzymes play a pivotal role in deinking recycled paper—a process that is increasingly important as the industry shifts toward circular economy models. Conventional deinking methods often struggle with the removal of ink, adhesives, and other contaminants, leading to lower fiber recovery and diminished pulp quality. Chimeric cellulase-xylanase constructs have shown enhanced performance in hydrolyzing ink particles and sticky residues, resulting in cleaner pulp and improved fiber yield. This enzymatic efficiency translates to lower energy consumption, reduced reliance on chemical solvents, and better alignment with global sustainability targets. As recycling becomes a cornerstone of modern papermaking, the role of chimeric enzymes in enabling high-quality secondary fiber processing will continue to grow.

Another critical application of chimeric enzymes is in pitch control. Pitch, a resinous substance naturally present in wood, can accumulate on machinery and interfere with paper formation, leading to defects and operational downtime. Traditional pitch control methods rely on chemical dispersants, which are costly and environmentally burdensome. Chimeric lipase-esterase enzymes offer a biological alternative by breaking down pitch components more effectively than their natural counterparts. These engineered enzymes exhibit broader substrate specificity and improved stability, allowing them to target a wider range of pitch compounds under industrial conditions. Studies have shown that their use reduces pitch-related deposits, enhances machine runnability, and improves the overall quality of paper products.

Beyond specific process enhancements, the integration of chimeric enzymes into pulp and paper manufacturing carries far-reaching economic, environmental, and operational implications. These multifunctional biocatalysts, engineered by fusing domains from different enzymes, offer a compelling alternative to conventional chemical treatments. By reducing the need for chlorine-based bleaching agents, alkalis, and other harsh reagents, chimeric enzymes help lower chemical consumption, minimize effluent toxicity, and decrease the overall environmental footprint of the industry. Additionally, their ability to function under extreme pH and temperature conditions allows for more efficient processing, which translates into reduced water usage, shorter reaction times, and lower energy requirements (Chaudhari et al. 2023). These improvements contribute directly to operational cost savings and a measurable reduction in greenhouse gas emissions.

The deployment of chimeric enzymes also enhances workplace safety by minimizing worker exposure to hazardous substances traditionally used in pulping and bleaching. As enzyme-based processes replace or supplement chemical-intensive steps, the risk of chemical burns, respiratory hazards, and long-term health effects is significantly reduced (Infinitabiotech 2023; Creative Enzymes 2023). Advances in synthetic biology, protein engineering, and fermentation technologies have further enabled the scalable and cost-effective production of chimeric enzymes. Modern expression systems, including genetically optimized microbial hosts, now support high-yield enzyme production with tailored post-translational modifications, making industrial deployment increasingly feasible (Bajpai 2018).

As enzyme engineering techniques continue to evolve, the customization of chimeric enzymes for specific substrates, feedstocks, and process conditions becomes more precise. This precision engineering allows for the development of enzyme variants with enhanced substrate affinity, improved thermostability, and resistance to inhibitors commonly found in lignocellulosic biomass. Such adaptability opens new avenues for innovation and industrial optimization, particularly in mills processing diverse raw materials such as hardwoods, softwoods, and agricultural residues (Chaudhari et al. 2023).

Recent research has also underscored the potential of chimeric enzymes in biorefinery applications, where lignocellulosic biomass is converted into value-added products such as biofuels, bioplastics, and platform chemicals. In this context, chimeric enzymes facilitate the selective depolymerization of cellulose, hemicellulose, and lignin, enabling efficient extraction of fermentable sugars and other intermediates critical for downstream fermentation and synthesis (Arsalan et al. 2025). Their dual utility—enhancing pulp and paper processes while supporting bio-based product development—positions them as key enablers of a more sustainable and diversified industrial ecosystem. Moreover, the modularity and high performance of chimeric enzymes make them suitable for integration into existing infrastructure with minimal retrofitting. This compatibility reduces barriers to adoption and accelerates the transition to greener technologies. Their use aligns with global sustainability goals and circular economy principles, particularly in regions aiming to reduce industrial emissions and improve resource efficiency.

Despite these advantages, the widespread adoption of chimeric enzymes in the pulp and paper sector is not without challenges. Regulatory approval processes for novel enzyme formulations can be time-consuming, and production costs—especially for highly engineered variants—may still be prohibitive for some facilities. Additionally, the need for industry-specific customization requires close collaboration between enzyme developers and mill operators. However, these barriers are being actively addressed through multi-stakeholder initiatives. Collaborative efforts involving academia, industry, and government agencies have led to pilot-scale demonstrations, funding support, and the development of knowledge-sharing platforms that facilitate technology transfer and best practices (Tanveer et al. 2023).

For example, several paper mills worldwide have successfully implemented enzyme-assisted pulping and bleaching protocols, demonstrating both the technical feasibility and the economic benefits of chimeric enzyme integration. These case studies serve as valuable models for other regions seeking to modernize their operations, reduce environmental impact, and enhance competitiveness in a resource-constrained global market.

Chimeric enzymes represent a paradigm shift in the pulp and paper industry (Arsalan et al. 2025). Their engineered capabilities offer targeted solutions to longstanding challenges related to efficiency, sustainability, and product quality. As research and development efforts continue to refine these biocatalysts, their role in shaping the future of papermaking is expected to expand significantly. By embracing chimeric enzyme technologies, the industry can move toward greener, more resilient manufacturing practices that align with global environmental and economic objectives.

The comparative advantages of chimeric enzymes over natural enzymes in the pulp and paper industry are presented in Table 1.1. Table 1.2 outlines the functional roles of chimeric enzymes across key process areas. Table 1.3 emphasizes the strategic benefits of integrating chimeric enzymes. In contrast, Table 1.4 examines the primary barriers to adoption—such as regulatory hurdles and production costs—alongside enabling factors like collaborative research, pilot-scale demonstrations, and advances in enzyme engineering that support broader implementation across the sector.

Table 1.1 Comparative benefits of chimeric versus. Natural enzymes in pulp and paper

Feature	Natural enzymes	Chimeric enzymes
Substrate specificity	Narrow	Broad/multifunctional
Thermal stability	Low to moderate	High
Customizability	Low	High (domain fusion)
Industrial tolerance	Limited	Robust under harsh conditions
Environmental impact	Moderate	Low (reduced chemical use)

Table 1.2 Applications of chimeric enzymes in pulp and paper processes

Process area	Chimeric enzyme type	Functional role	Benefits
Bleaching	Xylanase	Hemicellulose breakdown	Improved brightness, reduced toxicity
Pitch control	Lipase-Esterase	Resin degradation	Reduced deposits, smoother operations
Biorefinery	Cellulase, Hemicellulase	Biomass conversion	Enhanced sugar yield, biofuel production
Deinking	Cellulase-Xylanase	Ink/sticky removal	Cleaner pulp, better fiber recovery

Table 1.3 Strategic advantages of chimeric enzyme integration

Dimension	Impact
Economic	Lower operational costs, reduced chemical procurement
Environmental	Reduced effluents, lower GHG emissions
Technical	Enhanced process efficiency, better product quality
Safety	Minimized exposure to toxic chemicals
Scalability	Compatible with existing infrastructure, fermentable production

Table 1.4 Challenges and enablers for chimeric enzyme adoption

Challenge	Mitigation strategy
Enzyme incompatibility in cocktails	Use of multifunctional chimeric constructs
High production cost	Advances in fermentation and expression systems
Regulatory approval delays	Early engagement with regulatory bodies, standardized safety testing
Mill-specific customization needs	Collaborative R&D with enzyme suppliers and academic partners

References

Arsalan A, Ravikumar Y, Tang X, Cao Z, Zhao M, Sun W, Qi X (2025) Chimeric enzymes in the pulp and paper making industry: current developments. Biotechnol Adv 108530

Bajpai PK (2011) Emerging applications of enzymes for energy saving in pulp and paper industry. Ippta J 23:181–186

Bajpai P (2018) Biotechnology for pulp and paper processing. Springer, Singapore. https://doi.org/10.1007/978-981-10-7853-8_10

Chaudhari YB, Várnai A, Sørlie M, Horn SJ, Eijsink VG (2023) Engineering cellulases for conversion of lignocellulosic biomass. Protein Eng des Sel 36:1–12

Chaurasia S, Bhardwaj NK (2019) Biobleaching-an ecofriendly and environmental benign pulp bleaching technique: a review. J Carbohydr Chem 38:108–187

Creative Enzymes (2023) Application of enzymes in pulp and paper industry. https://www.creative-enzymes.com/resource/application-of-enzymes-in-pulp-and-paper-industry_64.html

García-Paz FD, del Moral S, Morales-Arrieta S, Ayala M, Treviño-Quintanilla LG, Olvera-Carranza C (2024) Multidomain chimeric enzymes as a promising alternative for biocatalysts improvement: a minireview. Mol Biol Rep 51(1):410. https://doi.org/10.1007/s11033-024-09332-9

Henríquez-Gallegos S, Albornoz-Palma G, Andrade A, Filgueira D, M´endez- Miranda A, Teixeira Mendonça R, Pereira M (2024) Effect of enzyme lignin oxidation by laccase on the enzymatic-mechanical production process of lignocellulose nanofibrils from mechanical pulp. Cellulose 31:3545–3560. https://doi.org/10.1007/s10570-024-05784-1

Infinitabiotech (2023) Pulp and paper enzymes for eco-friendly processing. https://infinitabiotech.com/blog/unlocking-innovation-exploring-the-latest-advances-in-pulp-and-paper-enzyme-technology/

Kumar H, Christopher LP (2017) Recent trends and developments in dissolving pulp production and application. Cellulose 24:2347–2365. https://doi.org/10.1007/s10570-017-1285-y

Lee AA, Gervasio ED, Hughes RO, Maalouf AA, Musso SA, Crisalli AM, Woolridge EM (2023) Alginate encapsulation stabilizes xylanase toward the laccase mediator system. Appl Biochem Biotechnol 195:3311–3326. https://doi.org/10.1007/s12010-022-04296-7

Ma Y, Lee C-J, Park J-S (2020) Strategies for optimizing the production of proteins and peptides with multiple disulfide bonds. Antibiotics 9:541. https://doi.org/10.3390/antibiotics9090541

Monica P, Ranjan R, Kapoor M (2024) Lignocellulose-degrading chimeras: emerging perspectives for catalytic aspects, stability, and industrial applications. Renew Sust Ener Rev 199:114425. https://doi.org/10.1016/j.rser.2024.114425

Onwosi CO, Ezugworie FN, Onyishi CL, Igbokwe VC (2024) Lignocellulose biomass pretreatment for efficient hydrolysis and biofuel production. In: Advances in biofuels production, optimization and applications. Elsevier, pp 1–19. https://doi.org/10.1016/B978-0-323-95076-3.00001-6

Paice MG, Jurasek L (1984) Peroxidase catalyzed color removal from bleach plant effluent. Biotechnol Bioeng 1984(26):477–480

Ribeiro RS, Pohlmann BC, Calado V, Bojorge N, Pereira N (2019) Production of nanocellulose by enzymatic hydrolysis: trends and challenges. Eng Life Sci 19:279–291

Shahid S, Batool S, Khaliq A, Ahmad S, Batool H, Sajjad M, Akhtar MW (2023) Improved catalytic efficiency of chimeric xylanase 10B from *Thermotoga petrophila* RKU1 and its synergy with cellulases. Enzym Microb Technol 166:110213. https://doi.org/10.1016/j.enzmictec.2023.110213

Sharma A, Balda S, Gupta N, Capalash N, Sharma P (2020) Enzyme cocktail: an opportunity for greener agro-pulp biobleaching in paper industry. J Clean Prod 271:122573

da Silva V, Amatto I, Gonsales da Rosa-Garzon N, de Oliveira A, Sim~oes F, Santiago F, da Silva P, Leite N, Raspante Martins J, Cabral H (2022) Enzyme engineering and its industrial applications. Biotechnol Appl Biochem 69:389–409. https://doi.org/10.1002/bab.2117

Söderholm P, Bergquist AK, Söderholm K (2019) Environmental regulation in the pulp and paper industry: impacts and challenges. Curr Forestry Rep 5:185–198. https://doi.org/10.1007/s40725-019-00097-0

Sun M, Wang Y, Shi L, Klemeš JJ (2018) Uncovering energy use, carbon emissions and environmental burdens of pulp and paper industry: a systematic review and meta-analysis. Renew Sustain Energy Rev 92:823–833

Tanveer A, Gupta S, Dwivedi S, Yadav K, Yadav S, Yadav D (2023) Innovations in papermaking using enzymatic intervention: an ecofriendly approach. Cellulose 30(11):7393–7425. https://doi.org/10.1007/s10570-023-05333-2

Wei S, Liu K, Ji X, Wang T, Wang R (2021) Application of enzyme technology in biopulping and biobleaching. Cellulose 28:10099–10116

Yakubu A, Saikia U, Vyas A (2019) Microbial enzymes and their application in pulp and paper industry. In: Yadav A, Singh S, Mishra S, Gupta A (eds) Recent advancement in white biotechnology through fungi. Fungal biology. Springer, Cham. https://doi.org/10.1007/978-3-030-25506-0_12

Yang M, Li J, Wang S, Zhao F, Zhang C, Zhang C, Han S (2023) Status and trends of enzyme cocktails for efficient and ecological production in the pulp and paper industry. J Clean Prod 418:138196

Chapter 2
Pulp and Paper Making Process

Abstract The pulp and paper industry remains a critical global sector, driven by demand for packaging, hygiene products, and sustainable fiber-based materials. Despite challenges such as raw material volatility, digital disruption, and stringent environmental regulations, the industry is advancing through innovation and circular-economy strategies. The manufacturing process—from raw material preparation and pulping to bleaching, papermaking, and finishing—has increasingly incorporated enzymatic and cleaner technologies to reduce chemical use, energy consumption, and environmental impact. Regional dynamics continue to shape production trends, with Asia–Pacific leading growth, Europe prioritizing low-carbon innovation, and North America modernizing its infrastructure. As technological advances, biorefinery integration, and sustainability-focused practices expand, the sector is positioned to play a pivotal role in the evolving global bioeconomy.

Keywords Pulping · Papermaking · Enzymatic processing · Sustainability · Circular economy · Bleaching technologies · Fiber resources · Packaging demand · Biorefinery integration · Environmental regulations

2.1 Global Status of the Pulp and Paper Industry

In 2025, the global pulp and paper industry is estimated to be valued at around USD 350–360 billion, underscoring its ongoing significance in packaging, hygiene, education, and industrial applications. Despite encountering structural changes, the sector is projected to achieve a compound annual growth rate (CAGR) of approximately 1.7–2.0% through 2030, fueled by shifting consumer preferences and sustainability requirements (Fortune Business Insights 2025; Maia Research 2023; Precedence Research 2025).

The industry is generally categorized into packaging paper, printing and writing paper, tissue and hygiene products, and specialty papers. Notably, packaging paper has emerged as the fastest-growing segment, driven by the rapid expansion of

P. Bajpai, *Chimeric Enzymes Showing Promise in Pulp and Paper Sector*,
SpringerBriefs in Molecular Science, https://doi.org/10.1007/978-3-032-18003-2_2

e-commerce and the global transition towards sustainable alternatives to plastic (Brickwork Ratings 2024; Reports and Data 2023).

Conversely, the printing and writing paper segment continues to experience a decline due to the rise of digital media and a decrease in demand for traditional publishing formats. However, tissue and hygiene products are witnessing stable growth, especially in emerging markets where urbanization and public health awareness are on the rise.

Specialty papers, which include those utilized in filtration, medical, and technical applications, remain niche yet strategically vital for innovation and diversification.

The geographic distribution of pulp and paper production and consumption highlights distinct regional dynamics. The Asia–Pacific region dominates the global market, with China and India leading due to their extensive manufacturing capabilities, increasing domestic demand, and strategic investments in fiber sourcing and mill modernization (Precedence Research 2025).

Europe, while grappling with mature market conditions, has established itself as a frontrunner in circular-economy practices, focusing on fiber recovery, recycling, and low-carbon production technologies. The regulatory frameworks and environmental objectives set by the European Union have expedited the implementation of sustainable practices throughout the continent (AFRY 2023). North America is persistently investing in automation, digitalization, and bio-based alternatives, with a focus on transforming legacy infrastructure into facilities that are ready for the future. The region's commitment to operational efficiency and product innovation enhances its competitiveness, even in the face of declining demand in traditional paper sectors (Fortune Business Insights 2025).

Latin America and Africa are emerging markets that hold considerable untapped potential. These areas are experiencing a rise in investments directed towards pulp plantations, mill construction, and infrastructure development, bolstered by favorable climatic conditions and increasing domestic consumption. Nevertheless, challenges such as logistics, policy stability, and the availability of skilled labor continue to hinder rapid growth (Precedence Research 2025).

Despite its resilience, the pulp and paper industry is confronted with a complex set of challenges in 2025. The volatility of raw materials remains a significant issue, as the prices of wood and recycled fiber fluctuate due to climate variability, geopolitical tensions, and changing trade policies (Maia Research 2023). The availability and cost of fiber have a direct effect on production economics and the stability of the supply chain. Environmental regulations are becoming more stringent across various jurisdictions, forcing mills to lower emissions, enhance effluent treatment, and implement cleaner technologies. Adhering to air and water quality standards, especially in regions with active environmental oversight, necessitates considerable capital investment and operational changes (AFRY 2023).

Digital disruption is continuously altering demand patterns, particularly in the printing and writing paper sector, where digital media and electronic communication have led to a significant decline in consumption. Furthermore, supply chain disruptions—intensified by post-pandemic recovery, fluctuations in energy prices, and transportation bottlenecks—present operational risks and require adaptive logistics strategies. North America is persistently investing in automation, digitalization, and bio-based alternatives, with a focus on transforming legacy infrastructure into facilities that are ready for the future. The region's commitment to operational efficiency and product innovation enhances its competitiveness, even in the face of declining demand in traditional paper sectors (Fortune Business Insights 2025).

Latin America and Africa are emerging markets that hold considerable untapped potential. These areas are experiencing a rise in investments directed towards pulp plantations, mill construction, and infrastructure development, bolstered by favorable climatic conditions and increasing domestic consumption. Nevertheless, challenges such as logistics, policy stability, and the availability of skilled labor continue to hinder rapid growth (Precedence Research 2025).

Despite its resilience, the pulp and paper industry is confronted with a complex set of challenges in 2025. The volatility of raw materials remains a significant issue, as the prices of wood and recycled fiber fluctuate due to climate variability, geopolitical tensions, and changing trade policies (Maia Research 2023). The availability and cost of fiber have a direct effect on production economics and the stability of the supply chain. Environmental regulations are becoming more stringent across various jurisdictions, forcing mills to lower emissions, enhance effluent treatment, and implement cleaner technologies. Adhering to air and water quality standards, especially in regions with active environmental oversight, necessitates considerable capital investment and operational changes (AFRY 2023).

Digital disruption is continuously altering demand patterns, particularly in the printing and writing paper sector, where digital media and electronic communication have led to a significant decline in consumption. Furthermore, supply chain disruptions—intensified by post-pandemic recovery, fluctuations in energy prices, and transportation bottlenecks—present operational risks and require adaptive logistics strategies (Fortune Business Insights 2025). These challenges highlight the necessity for agility, innovation, and regulatory alignment throughout the value chain.

Sustainability has evolved from being a secondary concern to becoming a primary strategic focus for the pulp and paper sector. Mills around the globe are embracing environmentally friendly practices to lessen their ecological footprint and improve resource efficiency. For example, enzyme-assisted pulping provides a biologically based alternative to traditional chemical processes, enhancing fiber quality while decreasing chemical usage and energy consumption (Reports and Data 2023). The implementation of water reuse and zero-liquid-discharge systems is on the rise to reduce freshwater consumption and effluent release, in line with global water stewardship objectives. The creation of bioproducts signifies a new wave of innovation, with lignin-based adhesives and cellulose nanofibers emerging as sustainable substitutes for petroleum-derived materials (AFRY 2023). These innovations not only broaden revenue opportunities but also strengthen the industry's contribution to the developing bioeconomy. Principles of the circular economy are being integrated into operational frameworks, with the incorporation of recycled fibers and waste-to-energy projects becoming commonplace in forward-thinking mills (Fortune Business Insights 2025). Such strategies diminish reliance on landfills, reduce carbon emissions, and generate value from by-products. The industry is transitioning from volume to value, with sustainability as the new currency, indicating a significant shift in how success is defined and attained within the pulp and paper industry.

2.2 Manufacturing Process

The pulp and paper manufacturing process is a complex, multi-stage industrial operation that transforms lignocellulosic biomass into a wide array of finished paper products. This transformation begins with the systematic preparation of raw materials, which includes debarking, chipping, and screening to ensure uniform feedstock quality. The primary raw materials are wood-based sources, predominantly softwood and hardwood logs, which differ in fiber length and strength characteristics. Softwoods, such as pine and spruce, provide long fibers that enhance tensile strength, while hardwoods like eucalyptus and birch contribute shorter fibers that improve smoothness and printability. In response to regional constraints on forest resources, non-wood fibers such as bagasse (a sugarcane byproduct), bamboo, kenaf, and straw are increasingly incorporated into the feedstock mix. These alternative fibers offer renewable and locally available options that support sustainable sourcing strategies (Bajpai 2021). Additionally, recycled fiber derived from post-consumer waste (e.g., office paper, cardboard) and industrial scrap is integrated into the process to promote resource efficiency and align with circular economy principles (Bajpai 2015, 2024; Cinquini 2025).

Pulping—the core conversion step—is achieved through mechanical, chemical, or semi-chemical methods, each tailored to specific product requirements (Bajpai 2018a, 2010; Biermann 1996; Smook 2016; Sixta 2006; Sixta et al. 2006; Gellerstedt 2009; Gellerstedt and Henriksson 2008). *Mechanical pulping* involves physically grinding wood chips into fibers using refiners or grinders. This method yields a high volume of pulp but retains most of the lignin, resulting in lower brightness and reduced strength. It is commonly used for products like newsprint and directories where cost efficiency and bulk are prioritized. *Chemical pulping*, particularly the kraft process, employs a cooking liquor composed of sodium hydroxide and sodium sulfide to dissolve lignin and separate cellulose fibers. The resulting pulp is strong, dark-colored, and well-suited for packaging grades and printing papers. The *sulfite process*, which operates under acidic conditions using sulfurous acid and bisulfite ions, produces lighter-colored pulp with smoother texture, ideal for fine writing and printing papers. *Semi-chemical pulping* combines mild chemical treatment with mechanical refining, striking a balance between yield and fiber quality. This method is widely used for manufacturing corrugated medium, which requires moderate strength and stiffness (Cinquini 2025).

The predominant pulping operations conducted to date, especially those that utilize chemicals, are not environmentally sustainable. Currently, novel pulping techniques are being implemented for the production of various grades of pulp that incorporate enzymes (Lin et al. 2018). These enzymatic pulping methods are employed in the preparation of dissolving pulp for the creation of fibrous materials. In the enzymatically facilitated pulping process, enzymes such as xylanase, cellulase, and hemicellulase are used to extract the pulp from the remaining lignocellulosic material and for post-treatment (Lin et al. 2018; Rashmi and Bhardwaj Nishi 2010; Yang et al. 2019). Reports indicate that the purity of pulp generated through enzymatic processes is superior to that achieved through conventional kraft pulping techniques.

Following pulping, the pulp undergoes *bleaching and washing* to remove residual lignin and enhance brightness. Bleaching sequences vary depending on environmental and product specifications (Bajpai 2012, 2013, 2018a). Elemental chlorine-free (ECF) bleaching utilizes chlorine dioxide, while totally chlorine-free (TCF) processes rely on oxidizing agents such as hydrogen peroxide, ozone, and peracetic acid. TCF methods are favored by environmentally conscious mills due to their reduced generation of adsorbable organic halides (AOX), which are persistent pollutants (Reeve 1996; Sixta et al. 2006). Washing stages, often integrated with counter-current flow systems, help recover chemicals and minimize effluent discharge.

The *papermaking phase* encompasses several precision-controlled operations (Bajpai 2018b, Sixta 2006b). First, stock preparation involves refining, blending, and dilution of pulp to achieve desired fiber characteristics. The pulp slurry is then delivered to the forming section, where it is distributed onto a moving wire mesh to form a continuous sheet. Water is drained by gravity and vacuum, and the wet web is transferred to the press section, where mechanical pressure removes additional moisture. The sheet then enters the drying section, where steam-heated cylinders evaporate residual water, stabilizing the paper structure. Finishing operations include calendaring for smoothness, cutting, and packaging. Advanced mills integrate *enzymatic treatments* to enhance fiber bonding and reduce energy consumption during refining. *Surface sizing* with starch or synthetic polymers improves sheet strength and printability, while *functional coatings*—such as barrier layers for moisture, grease, or oxygen—enable the production of specialty papers for food packaging, medical applications, and electronics (StartUs Insights 2025).

The different pulping operations and the integrated paper manufacturing process are shown in Fig. 2.1.

Environmental stewardship is a critical dimension of pulp and paper manufacturing, governed by stringent regulatory frameworks across jurisdictions. The industry generates various emissions and effluents, including particulate matter and sulfur compounds from combustion processes, as well as chemical oxygen demand (COD), biological oxygen demand (BOD), and suspended solids in wastewater streams. Solid waste such as sludge, bark, and ash must be managed through landfill diversion, composting, or energy recovery. Regulatory compliance is enforced through national and international standards. In India, the Central Pollution Control Board (CPCB) mandates effluent discharge norms and air quality limits for paper mills. The European Union applies the Integrated Pollution Prevention and Control (IPPC) Directive, emphasizing Best Available Techniques (BAT) for minimizing environmental impact. In the United States, the Environmental Protection Agency (EPA) enforces the Cluster Rule, integrating air and water regulations for pulp mills.

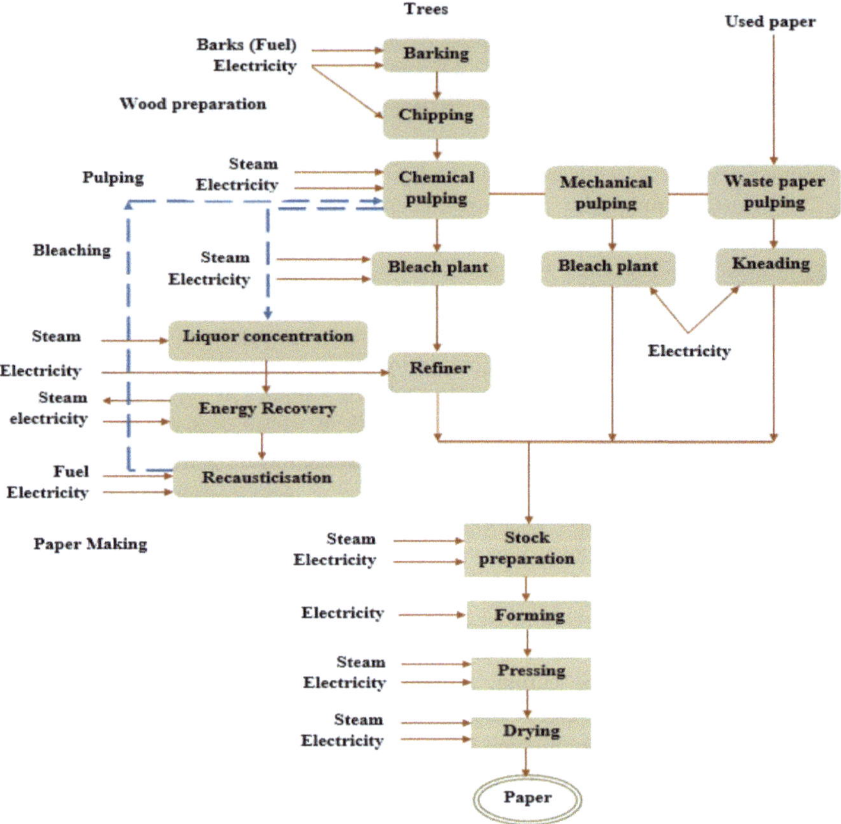

Fig. 2.1 Process flow in Pulp and Paper manufacturing mills. Haile et al. (2021). Licensed under a Creative Commons Attribution 4.0 International License

Sustainable certifications play a pivotal role in validating environmental performance and responsible sourcing. The Forest Stewardship Council (FSC) and Programme for the Endorsement of Forest Certification (PEFC) ensure that fiber is harvested from sustainably managed forests. Ecolabels such as the Blue Angel and Nordic Swan further recognize mills that meet rigorous environmental criteria, including energy efficiency, chemical safety, and recyclability. These frameworks not only support regulatory compliance but also enhance market access and consumer trust in eco-labeled products.

Technological innovation is reshaping the pulp and paper industry, enabling greater efficiency, sustainability, and product diversification. Digitalization is at the forefront, with artificial intelligence (AI) and machine learning deployed for quality control, predictive maintenance, and process optimization. Digital twins—virtual replicas of physical assets—allow real-time monitoring and simulation of mill operations, reducing downtime and enhancing decision-making. Automation and robotics are increasingly used in material handling, inspection, and packaging, improving safety and throughput.

Biorefinery integration represents a transformative shift, positioning pulp mills as platforms for producing bioethanol, bioplastics, and nanocellulose. Valorization of lignin and hemicellulose, traditionally considered waste, enables the creation of high-value chemicals and materials. These innovations align with the principles of the bioeconomy, reducing reliance on fossil resources and expanding the industry's role in sustainable development. Carbon neutrality goals are also driving investment in renewable energy sources such as biomass and solar, as well as carbon capture and storage (CCS) technologies. Comprehensive tracking of Scope 1, 2, and 3 emissions is becoming standard practice, supported by digital platforms and regulatory reporting tools.

Consumer preferences and market dynamics are evolving rapidly, influencing product development and strategic priorities within the pulp and paper sector. The packaging revolution is a defining trend, with increasing demand for paper-based alternatives to plastic in food, retail, and logistics applications. Molded fiber products and barrier-coated papers are gaining popularity due to their compostability and recyclability. Hygiene and tissue products have experienced sustained growth, particularly in urban markets where public health awareness and disposable income are rising. Premium tissue offerings, including lotion-infused and antimicrobial variants, reflect consumer demand for comfort and functionality.

Conversely, the print media segment continues to decline, driven by digital substitution and changing communication habits. While demand for office paper and newsprint has contracted, niche markets for archival, security, and specialty papers remain viable. These include watermarked, anti-counterfeit, and high-opacity papers used in legal, financial, and artistic applications. The industry's ability to adapt to these trends through product innovation and market segmentation is essential for long-term resilience.

Looking ahead, the pulp and paper industry is poised for cautious growth, shaped by sustainability imperatives, technological progress, and shifting global dynamics. Key growth drivers include the expansion of sustainable packaging, the rise of bio-based innovations, and supportive regulatory incentives. Governments and financial institutions are increasingly offering green financing and subsidies for eco-friendly infrastructure, encouraging mills to modernize and decarbonize. Public–private partnerships and academic-industry collaborations are also fostering research and development in fiber science, enzymology, and process engineering. However, risks persist. Climate-induced raw material shortages, trade barriers, and geopolitical tensions could disrupt supply chains and investment flows. The industry must navigate these uncertainties through strategic planning, diversification, and stakeholder engagement. Opportunities lie in leveraging digital tools for transparency, adopting circular economy models, and aligning with global sustainability frameworks such as the United Nations Sustainable Development Goals (SDGs). With proactive leadership and adaptive capacity, the pulp and paper sector can evolve into a cornerstone of the bioeconomy, delivering value across environmental, economic, and social dimensions.

References

AFRY (2023) World paper markets up to 2035: outlook for demand, supply, and trade. AFRY Management Consulting. https://afry.com/sites/default/files/2023-09/world_paper_market_up_to_2035_2023_edit_a4_final_0.pdf

Bajpai P (2012) Environmentally benign approaches for pulp bleaching, 2nd ed. Elsevier

Bajpai P (2010) Environmentally friendly production of pulp and paper. John Wiley & Sons

Bajpai P (2015) Green chemistry and sustainability in pulp and paper industry. Springer

Bajpai P (2018a) Biermann's handbook of pulp and paper: volume 1, raw material and pulp making, 3rd ed. Elsevier

Bajpai P (2018b) Biermann's handbook of pulp and paper: volume 2: paper and board making, 3rd ed. Elsevier

Bajpai P (2021) Nonwood plant fibers for pulp and paper. Elsevier. https://doi.org/10.1016/C2019-0-02638-2

Bajpai P (2024) Recycling and deinking of recovered paper, 2nd edn. Elsevier, Netherlands

Biermann CJ (1996) Handbook of pulping and papermaking, 2nd ed. Academic Press

Brickwork Ratings (2024) Pulp & paper industry: sector update 2024. Brickwork Ratings India Private Limited. https://www.brickworkratings.com/Research/Pulp%20%26%20Paper%20Industry_Final.pdf

Cinquini S (2025) Processes, tech, and sustainability should be 2025 focus. https://paper360.tappi.org/2025/06/18/processes-tech-and-sustainability-should-be-2025-focus/

Fortune Business Insights (2025) Pulp and paper market size, share, growth and industry forecast, 2025–2035. https://www.fortunebusinessinsights.com/pulp-and-paper-market-103447

Gellerstedt G, Henriksson G (2008) Wood chemistry and biotechnology. In Ek M, Gellerstedt G, Henriksson G (eds) Pulp and paper chemistry and technology, vol 1. De Gruyter

Gellerstedt G (2009) Chemistry of bleaching of wood pulps. In: Ek M, Gellerstedt G, Henriksson G (eds) Pulp and paper chemistry and technology, vol 2. Walter de Gruyter, pp 91–120

Haile A, Gelebo GG, Tesfaye T, Mengie W, Mebrate MA, Abuhay A, Limeneh DY (2021) Pulp and paper mill wastes: utilizations and prospects for high value-added biomaterials. Bioresour Bioprocess 8:35. https://doi.org/10.1186/s40643-021-00385-3

Lin X, Wu Z, Zhang C, Liu S, Nie S (2018) Enzymatic pulping of lignocellulosic biomass. Ind Crops Prod 120:16–24

Maia Research (2023) Global pulp and paper market insights 2023–2030. https://www.maiaresearch.com/Press_Release/1864139.html

Precedence Research (2025) Pulp and paper market–global industry analysis and forecast 2030. https://www.precedenceresearch.com/pulp-and-paper-market

Rashmi S, Bhardwaj Nishi K (2010) Enzymatic refining of pulps: an overview. IPPTA J 22:109–116

Reeve DW (1996) Introduction to the principles and practice of pulp bleaching. In: Dence CW, Reeve DW (eds) Pulp bleaching: principles and practice. TAPPI Press, pp 1–25

Reports and Data (2023) Paper and pulp market–global outlook 2023–2030. https://www.reportsanddata.com/report-detail/paper-and-pulp-market

Sixta H (ed) (2006) Handbook of pulp, vol 1. Wiley-VCH

Sixta H, Süss HU, Potthast A, Schwanninger M, Krotscheck AW (2006) Pulp bleaching. In: Sixta H (ed) Handbook of pulp. Wiley-VCH, pp 609–932. https://doi.org/10.1002/9783527619887.ch7a

Smook GA (2016) Handbook for pulp & paper technologists, 4th ed. TAPPI Press

Yang S, Yang B, Duan C, Fuller DA, Wang X, Chowdhury SP, Stavik J, Zhang H (2019) Applications of enzymatic technologies to the production of high-quality dissolving pulp XE "Dissolving pulp": a review. Biores Technol 281:440–448

Chapter 3
Utilization of Enzymes and Enzyme Blends Within the Pulp and Paper Sector

Abstract This chapter provides a comprehensive overview of the diverse enzymatic strategies employed in pulp and paper processing, emphasizing their growing role in sustainable and energy-efficient manufacturing. It highlights key classes of enzymes—including xylanases, laccases, pectinases, mannanases, esterases, and lignin-degrading oxidoreductases—and explains their specific functions in biopulping, biobleaching, refining, sticky control, and deinking. The chapter underscores the advantages of enzyme cocktails over single-enzyme systems, particularly their synergistic ability to enhance delignification, improve brightness, strengthen fiber properties, and reduce chemical consumption. Influencing factors such as pH, temperature, pulp type, and enzyme–substrate interactions are discussed to demonstrate the importance of process optimization. Collectively, the chapter presents enzymes as essential tools for advancing greener, more efficient pulp and paper technologies.

Keywords Enzyme cocktails · Biobleaching · Biopulping · Xylanase · Laccase · Sticky control · Deinking · Refining enzymes · Lignin degradation · Pulp and Paper biotechnology

In recent years, enzyme applications in the pulp and paper industry have gained momentum. Integrating biological treatments with milder conventional methods offers promising solutions to challenges in existing processes. Enzymes are increasingly recognized for boosting productivity, minimizing environmental impact, and lowering energy demands (Bajpai 2011).

The first documented use of enzymes in the pulp and paper industry dates back to 1984, marking a shift toward biotechnological innovation in fiber processing (Paice and Jurasek 1984). Early enzyme systems faced limitations due to high costs, poor recyclability, and low stability under industrial conditions, hindering broad adoption.

Advances in enzyme engineering and industrial biotechnology have since improved enzyme robustness, efficiency, and cost-effectiveness. These developments have enabled the commercialization of diverse enzymatic treatments now routinely applied across multiple stages of pulp and paper production. Enzyme applications in pulp and paper operations are depicted in Fig. 3.1.

P. Bajpai, *Chimeric Enzymes Showing Promise in Pulp and Paper Sector*, SpringerBriefs in Molecular Science, https://doi.org/10.1007/978-3-032-18003-2_3

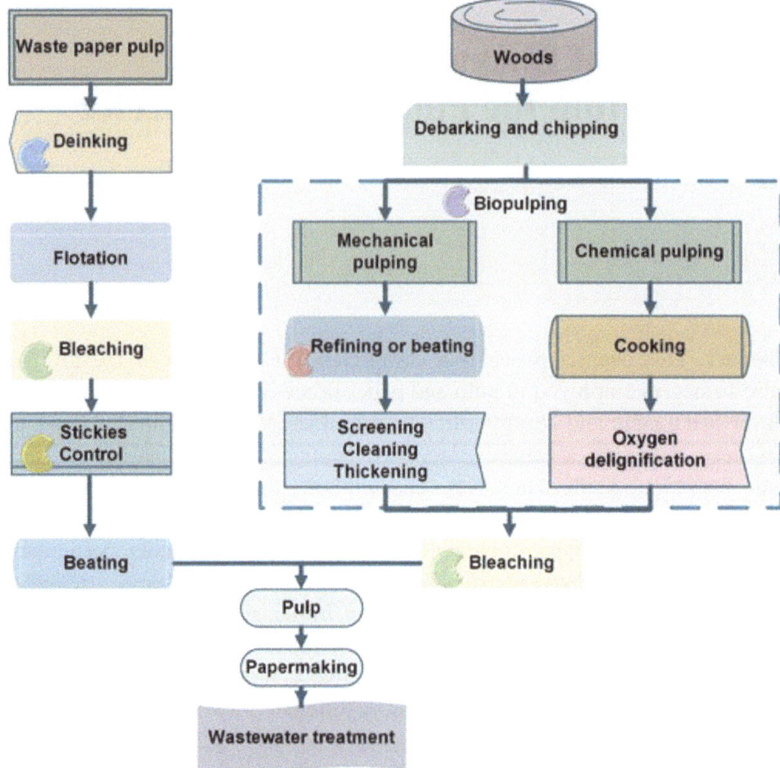

Fig. 3.1 Representation of enzyme applications within P&P processes. The figure's notes highlight the moments in the process when enzymes can be added. Yang et al. (2023). Reproduced with permission

Enzymes are employed to enhance operations such as fiber modification, pitch control, bleaching, deinking, and drainage. Their use has led to improved pulp yield, fiber quality, and paper strength, while reducing chemical inputs and energy consumption—supporting the industry's sustainability goals.

Emerging areas like dissolving pulp and nanocellulose production show strong potential for enzymatic innovation. Tailored enzyme formulations in these domains offer precise fiber modification, higher purity, and better functional properties (Kumar and Christopher 2017; Ribeiro et al. 2019).

The enzyme market for pulp and paper industry is rapidly evolving, with global biotech firms delivering scalable, compliant platforms. This reflects the strategic role of enzyme-enabled processes in advancing eco-friendly and economically viable technologies. However, enzymatic performance in the pulp and paper industry is shaped by a complex matrix of variables—enzyme dosage temperature, reaction pH residence time and pulp consistency—all requiring careful optimization for consistent results.

Enzyme-substrate interactions are particularly critical. Variability in pulp fiber sources, chemical additives, and recycled paper impurities can significantly affect enzyme accessibility, specificity, and catalytic efficiency (Immerzeel and Fiskari 2023; Bajpai 2018; Wei et al. 2021; Chen et al. 2018).

No single enzyme can address all operational challenges across unit processes. This has led to the development of customized enzyme blends—"cocktails"—designed to harness synergistic activities for targeted performance, reduced chemical reliance, and enhanced sustainability.

Formulating effective enzyme cocktails is now a key focus in industrial biotechnology, with ongoing research aimed at improving compatibility, stability, and integration within complex PPI workflows. This approach supports both process optimization and compliance with environmental regulations.

3.1 Enzymes and Supporting Proteins in Pulp and Paper Processing

Enzyme selection in pulp and paper processes depends on the specific substrates and operational goals. Table 3.1 outlines commonly used enzymes, their substrates, and functions. Hydrolases (e.g., cellulases, hemicellulases, pectinases, lipolytic enzymes, amylases) and oxidoreductases (e.g., lignin-degrading enzymes, lytic polysaccharide monooxygenases or LPMOs) are frequently employed in the pulp and paper industry workflows (Wei et al. 2021; Sharma et al. 2021; Bajpai 1999). Given the compositional complexity of the pulp and paper industry environments, enzyme targets extend beyond lignin. Specialized enzymes are often required: lipolytic enzymes address stickies from vegetable fibers and additives in recycled paper, while amylases facilitate deinking by breaking down coatings (Liu et al. 2018; Akbarpour and Resalati 2022).

Lignin-Degrading Enzymes

Lignin degradation is achieved through enzymatic and nonenzymatic means. Key enzymes include lignin peroxidase (LiP, EC 1.11.1.14), phenol oxidases (EC 1.10.3.2), versatile peroxidase (VP, EC 1.11.1.16) and manganese peroxidase (MnP, EC 1.11.1.13) (Kamimura et al. 2019).

Table 3.1 Enzymes used in pulp and paper processing

Cellulases—Endoglucanase (EC 3.2.1.4), Exoglucanase (EC 3.2.1.91), β-Glucosidase (EC 3.2.1.21) Substrate is Cellulose Utilized in Refining/beating, Biobleaching, Biodeinking Their roles are in Removal of the primary wall and S1 layer, increasing the specific surface area of single fibers; promoting the peeling of ink particles from the fiber surface layer (Gupta et al. 2015a; Peng et al. 2018; Tyagi et al. 2011; Zhang et al. 2013)
Cellulose synergistic proteins—Carbohydrate-binding modules, Swollenin, Lytic polysaccharide monooxygenase Substrate is Cellulose Utilized in Refining/beating Their roles are in Synergistic promotion of cellulase hydrolysis (Cebreiros et al. 2021; Guo et al. 2017; Hu et al. 2018; Ravalason et al. 2009, 2012)
Xylanase—Endo-1,4-β D-xylanase (EC 3.2.1.8), β-D-Xylosidase (EC 3.2.1.37), α-L-Arabinofuranosidases (EC 3.2.1.55), α-D-Glucuronidases (EC 3.2.1.139) Substrate is Xylan Utilized in Biopulping, Refining/beating, Biobleaching Their roles are in Xylan degradation to assist in lignin degradation; improvement of fiber surface permeability and degradation of xylan derivatives containing chromogenic groups to assist in bleaching (Khambhaty et al. 2018; Liu et al. 2012; Nagar and Gupta 2021; Wu et al. 2018; Zheng et al. 2012)
Pectinase—Polygalactronase (EC 3.2.1.15), Pectate lyase (EC 4.2.2.2), Pectin lyase (EC 4.2.2.10) Substrates are Pectic substances Their roles are in Biopulping, Biobleaching, Sticky control, Pectin degradation to form monofibrils for improved pulp yield and paper brightness; polygalacturonic acid degradation to reduce sticky compound deposition (Ahlawat et al. 2008; Jiang et al. 2018; Li et al. 2013b)
Lipolytic enzymes—Lipases (EC 3.1.1.3), Cutinase (EC 3.1.1.74), Esterases (EC 3.1.1.13) Substrates are Triglycerides, Fatty acids, Esters Utilized in Biobleaching, Sticky control Their roles are in Ester hydrolysis to remove ink from fibers and reduce sticky compound adhesion (Ballinas-Casarrubias et al. 2020; Cedillo et al. 2013; Liu et al. 2018; Nathan and Rani 2020; Yang et al. 2018; Yin et al. 2021)
Amylase—α-Amylases (EC 3.2.1.1) Substrate is Starch Utilized in Biodeinking, Sticky control Their roles are in Starch hydrolysis on the ink surface to aid deinking (Akbarpour and Resalati 2022; Ali et al. 2014; Ebrahimi et al. 2014; Gao et al. 2018)

(continued)

Table 3.1 (continued)

Lignin degrading enzymes—
Laccases (EC 1.10.3.2), Lignin peroxidase (EC 1.11.1.14), Manganese peroxidase (EC 1.11.1.13), Versatile peroxidase (EC 1.11.1.16)
Substrate is lignin
Utilized in Lignin Biopulping, Refining/beating, Biobleaching processes
Their roles are in Lignin degradation to improve paper brightness and strength (Chen et al. 2012; Jha and Patil 2013; Li et al. 2013a; Xu et al. 2010)
Lignin auxiliary enzymes –
Aryl-alcohol oxidase (EC 1.1.3.7), Alcohol oxidase (EC 1.1.3.13), Glyoxal oxidase (EC 1.2.3.15), Glucose oxidase (EC 1.1.3.4), Pyranose oxidase (EC 1.1.3.10), Aryl-alcohol dehydrogenases (EC 1.1.1.90), Cellobiose dehydrogenase (EC 1.1.99.18)
Substrate is lignin
Utilized in lignin, Biopulping, Refining/beating
Their roles are in Enhancement of lignin degradation to reduce pulp yellowness and improve pulp whiteness (Fang et al. 1999; Flitsch et al. 2013; Sahinkaya et al. 2019; Schou et al. 1994)

Based on Bajpai (2018)

Laccases (EC 1.10.3.2), the most widely applied, function in pulping, refining, bleaching, and deinking. These multicopper oxidases cleave lignin via Cα-oxidation and aryl-alkyl bond disruption, enhancing paper brightness and strength—often matching or surpassing chemical treatments. Alone, laccases oxidize phenolic lignin units; with redox mediators, they also target nonphenolic compounds. This synergistic approach, known as the laccase mediator system (LMS), employs potent mediators like 1-HBT, NHA, and VA (Rodrigues et al. 2019; Chauhan et al. 2017; Woolridge 2014; Singh and Arya 2019; Bajpai 2018). Despite promising results in biopulping and biobleaching, LMS drawbacks include reduced polymerization, cellulose oxidation via hydroxyl radicals, and mediator cost/toxicity. Laboratory pretreatment of OCC pulp with Lac/Gly improved whiteness and brightness by 16.17% and 7.41%, respectively, and increased carboxyl content (Chen et al. 2021).

Manganese peroxidases oxidize Mn^{2+} to Mn^{3+} and phenolic compounds using H_2O_2, reducing pulp kappa values and enhancing bleachability (Dashtban et al. 2010; Woolridge 2014). However, MnPs face commercial limitations due to low availability, peroxide sensitivity, and residual Mn^{2+} effects on whiteness. Recombinant techniques are being explored to address these issues (Xu et al. 2010). A lab-scale study combining MnP with laccase, glucose oxidase, and additives achieved 65.85% lignin degradation in maize straw within three days (Chen et al. 2022).

Lignin peroxidases, which are glycoproteins containing heme, facilitate the H_2O_2-dependent depolymerization of lignin, both phenolic and nonphenolic, with or without mediators (Dashtban et al. 2010).

Effective biopulping of oil palm empty fruit bunches by employing LiP and xylanase, modifying the lignin-hemicellulose interface to generate cellulose pulp in an eco-friendly manner was illustrated by Colonia et al. (2019)

Versatile peroxidases are hybrid enzymes with multiple substrate binding sites, enabling oxidation of Mn^{2+} and diverse lignin compounds. Their broader oxidation capabilities surpass those of LiPs and MnPs (Dashtban et al. 2010). In lab-scale trials, lignin in paddy straw and pine needles was degraded using fungi that produce LMEs—laccase, MnP, and LiP—combined with mild chemical treatments (Gupta et al. 2023).

Cellulases

Cellulases—endo-(1,4)-β-D-glucanase, exo-(1,4)-β-D-glucanase, and β-glucosidase work synergistically to hydrolyze cellulose into glucose. EG cleaves β-1,4-glycosidic bonds in both amorphous and crystalline regions, reducing polymerization; CBH removes glucose units from the nonreducing end, producing cellobiose; BG converts cellooligosaccharides and cellobiose into glucose but does not act on crystalline cellulose (Singh et al. 2016b; Kuhad et al. 2011).

Primarily used in refining/beating, cellulases also support pulping, deinking, and bleaching. In pulping, they remove primary and outer secondary walls to expose the inner secondary wall, increasing fiber surface area and reducing energy use (Kumar et al. 2021a; Chen et al. 2014). During refining, they enhance fiber bonding and cohesion. Fungal sources such as *Thermomonospora, Trichoderma,* and *Aspergillus* are under active investigation (Abdulhadi and Ashish 2021). Studies confirm cellulases improve pulp drainage, physical properties, and reduce energy consumption (Peng et al. 2018; Lee et al. 2020).

Xylanases

Xylanases, a class of hemicellulases, include endo-1,4-β-D-xylanase, β-D-xylosidase, α-L-arabinofuranosidase, α-D-glucuronidase, and acetylxylan esterase. Endo-xylanases are most prevalent in PPI (Collins et al. 2005). Since 1984, xylanases have been widely used in bleaching, pulping, and deinking (Paice and Jurasek 1984; Singh et al. 2016a; Bajpai 2018).

They hydrolyze lignin-carbohydrate complexes, loosen cell walls, and remove chromophores, enhancing chemical diffusion and lignin degradation, thereby improving brightness (Clark et al. 1991; Woolridge 2014). Pretreatment reduces chlorine use and VOC emissions, improving fibrillation and reducing effluent toxicity (Dutta et al. 2020; Singh et al. 2019). Commercial xylanases, mostly fungal and belonging to family 11, are increasingly purified or genetically modified to avoid cellulase-related drawbacks. Examples include Pulpzyme HA/C, Novozyme 473, and Cartazymes HS-10.

Pectinases

Pectin, a complex polysaccharide, binds with cellulose, hemicellulose, and lignin to form plant cell structures; in pulp, it contributes to yellowing (Satapathy et al. 2020). Pectinases—such as pectate lyases and polygalacturonases—are used in pulping and sticky control. They degrade pectin in bast fibers, improving fiber separation and pulp yield (Nawawi et al. 2021).

Pectinases also reduce sticky deposits and break down polygalacturonic acid in dissolved and colloidal substances, enhancing cationic polymer efficiency (Li et al. 2013b). Microorganisms, especially fungi, are the main commercial sources

(Hugouvieux-Cotte-Pattat et al. 2014). Products include Pectinex™ and Pectinex SP-L. Lab studies show improved whiteness, reduced kappa value, and lower cationic demand with pectinase treatment (Jiang et al. 2018).

Lipolytic Enzymes

Lipolytic enzymes—lipases, cutinases, and esterases—belong to the α/β-hydrolase superfamily and share a Ser-Asp/Glu-His catalytic triad (Gricajeva et al. 2022). They are widely used in deinking and sticky control. Lipases hydrolyze long-chain acyl esters, esterases act on short-chain soluble esters, and cutinases target medium-chain esters and synthetic binders (Cedillo et al. 2013).

Lipases degrade triglycerides found in inks and adhesives, reducing viscosity and aiding ink removal. Byproducts act as surfactants to prevent stickies from reagglomerating (Ballinas-Casarrubias et al. 2020). Combined use with cellulase and xylanase achieved 14.5% brightness in deinked pulp (Nathan and Rani 2020).

Cutinases hydrolyze polyesters and binders like polyvinyl acetate. A chimeric lipase-cutinase (Lip-Cut) improved enzymatic activity by over 100% (Liu et al. 2018).

Esterases include carboxylesterases and sterol esterases. They break ester bonds in adhesives, reducing sticky deposits. A novel esterase, DacApva, deacetylates PVA for sticky control (Yin et al. 2021). Feruloyl esterase from *A. alternata* expressed in *Pichia pastoris* enhanced dissolving pulp reactivity when combined with endoglucanase (Hou et al. 2022).

Amylases

α-Amylases hydrolyze glycosidic bonds in starch to produce dextrins and smaller polymers (Janeček et al. 2014). In deinking, they break down starch coatings, facilitating ink removal and improving cellulose hydration. Amylases are often used with xylanase and lipase for wastepaper deinking. One study reported 19.9% deinking efficiency and improved tensile and tear indices (Gao et al. 2018). Lab-scale refining of OCC pulp with amylase showed enhanced dehydration and physical properties compared to untreated pulp (Akbarpour and Resalati 2022).

Other essential enzymes and proteins

Beyond ligninases and cellulases, various natural enzymes and intracellular proteins enhance substrate breakdown in pulp and paper processes. These auxiliary components synergize with hydrolytic enzymes to improve efficiency, reduce reaction time, and are increasingly integrated into commercial multienzyme formulations.

Extracellular fungal oxidases—such as AAO, alcohol oxidase, GLOX, GOx, and P2O—and dehydrogenases like AADs and CDH generate reactive species (e.g., H_2O_2) that facilitate lignin cleavage by creating access points in plant cell walls. For instance, a peroxidase from *Rhodococcus* sp. T1 reduced kappa number by 5.2 units and increased delignification by 12% during eucalyptus kraft pulp bleaching.

Nonhydrolytic proteins such as carbohydrate-binding modules (CBMs), expansins, and lytic polysaccharide monooxygenases (LPMOs) further support lignocellulose hydrolysis. CBMs anchor enzymes to cellulose surfaces, enhancing local enzyme concentration and substrate degradation. Expansins—including fungal analogs like swollenin and loosenin—disrupt cellulose crystallinity and boost cellulase activity by up to 240%. LPMOs (AA9 family), copper-dependent enzymes, oxidize glucose units to expose new chain ends, significantly increasing hydrolysis yields under aerobic conditions.

Commercial enzyme blends such as Cellic Ctec2/3 (Novozymes) incorporate AA9 and outperform earlier formulations. Coexpression of xylanase, LPMO, and SWO in *Yarrowia lipolytica* enhanced degradation of crystalline cellulose and industrial pulp. Additionally, LPMO and CBM combinations accelerated nanofibrillation of mechanical pulp in lab-scale studies.

3.2 Enzyme Cocktails in Pulp and Paper Processing

The effectiveness of enzyme cocktails varies across different papermaking operations, necessitating tailored formulations for each process. For example, in biopulping, excessive xylanase can impair pulp quality, whereas optimal dosing enhances lignin removal and improves brightness. In biodeinking, ink particle size and wastepaper type significantly influence enzymatic efficiency, which can be optimized by adjusting enzyme type, concentration, reaction time, and incorporating auxiliary proteins (Kumar et al. 2021b).

Sticky control presents additional challenges, as certain stickies—particularly those that are fibrous and hydrophobic—impede enzyme-substrate binding. Compounds like alkyl succinic anhydride form covalent bonds with cellulose, increasing hydrophobicity and reducing enzymatic activity (Kumar et al. 2021b). Enzyme selection must therefore align with process-specific requirements; while some enzymes are versatile across multiple applications, others are highly specialized (Colonia et al. 2019). Developing effective enzyme cocktails requires a comprehensive understanding of the operational parameters and substrate characteristics unique to each papermaking stage.

Biopulping

Traditional pulping is one of the most energy-intensive operations in paper manufacturing. Biopulping—defined as the enzymatic or microbial pretreatment of raw materials prior to conventional pulping—offers a sustainable alternative. It encompasses biochemical and biomechanical approaches, and has been shown to improve pulp yield and strength, reduce chemical consumption during cooking and bleaching, and lower energy use and effluent pollution (Nagpal et al. 2021; Kumar et al. 2020; Jha and Patil 2013; Wei et al. 2021; Walia et al. 2015).

Microbial pulping involves the direct application of lignin-degrading fungi such as white rot, brown rot, and soft rot species (Bajpai 2018). Despite its potential, microbial pulping remains limited due to equipment constraints and process control challenges. Enzymatic pulping, while slightly more costly, is preferred for its operational simplicity and reliability.

Biopulping enzymes include hemicellulases, ligninases, and pectinases. Tethys Research LLC is developing hemicellulose:lignin etherases (HLEs)—a class of enzymes including mannanases and xylanases—that selectively cleave lignin-hemicellulose linkages without degrading the substrate. Laccase directly depolymerizes lignin, while xylanase disrupts lignin-hemicellulose bonds, enhancing delignification (Woolridge 2014).

Laboratory studies confirm the superior performance of enzyme cocktails over single enzymes or chemical methods. For instance, laccase-xylanase treatment produced adhesive-free bamboo boards with improved mechanical properties compared to laccase-only pretreatment. Biopulping of rice straw using a xylanase:pectinase ratio of 400:120 IU/g enhanced paper quality, yield, and brightness while reducing alkali requirements. Similarly, oil palm empty fruit bunches treated with xylanase and lignin peroxidase (LiP) achieved a pulp yield of 74.36%, with higher lignin and hemicellulose removal when biopulping was combined with biobleaching versus alkaline-biobleaching methods (Song et al. 2018; Colonia et al. 2019; Nagpal et al. 2021).

Refining and Beating

Enzymatic refining/beating involves the controlled degradation and softening of fibers to activate the fiber surface and facilitate cell wall delamination. This process significantly reduces energy consumption while enhancing fiber disintegration, swelling, fibrillation, drainage, blending properties, and overall paper strength

Cellulases are the primary enzymes used, often supplemented by xylanase and laccase. Novozymes' NS-22086, a multienzyme formulation containing cellulase, xylanase, and other glycosidic hydrolases, has demonstrated efficacy in pretreating pine kraft and other cellulosic pulps. DuPont's AccelleraseRTRIO™—comprising CBH, EG, xylanase, and β-glucosidase—has shown enhanced hydrolysis of wheat straw and hardwood pulp (Torres et al. 2012; Taneda et al. 2012; Smit and Huijgen 2017; Buzała et al. 2015, 2016, 2017).

Engineered strains co-producing alkali-resistant EG and xylanase have improved drainage and strength properties in wheat straw pulp compared to single-enzyme treatments. Mixed enzyme treatments outperform single-enzyme applications in enhancing bleached pine pulp strength. Process sequencing also influences outcomes; cellulase and xylanase may exhibit synergistic, antagonistic, or neutral effects depending on their order of addition (Wang et al. 2017; Cui et al. 2016; Buzała et al. 2016).

Additionally, LPMOs have been employed in kraft pulp refining to promote nanofibrillation, yielding a more uniform nanofiber composition without altering core fiber properties (Hu et al. 2018).

Biobleaching

Biobleaching uses enzymes to reduce chemical load and improve pulp brightness. Key enzymes include cellulases, xylanases, and laccases, which break down fiber structures and enhance chemical penetration. Commercial blends like Novozymes' NS-22086 and DuPont's AccelleraseRTRIO™—containing CBH, EG, xylanase, and β-glucosidase—have shown improved hydrolysis of various pulps. Engineered strains co-producing alkali-resistant EG and xylanase enhanced drainage and strength in wheat straw pulp. Mixed enzyme treatments outperform single-enzyme applications in refining bleached pine pulp, and LPMOs promote nanofibrillation without altering fiber integrity (Hu et al. 2018; Buzała et al. 2015, 2016; Smit and Huijgen 2017; Wang et al. 2017). Table 3.2 illustrates the various enzyme applications in biobleaching that have been developed in recent years.

Numerous laboratory-scale investigations have been carried out to elucidate the complex interactions between enzymes and various paper-processing additives, as outlined in Table 3.3. These studies collectively highlight how additive selection significantly influences enzymatic performance and pulp quality. For instance, Nagpal et al. (2021) examined the effects of different wetting agents on xylanase and pectinolytic enzymes during the biopulping of rice straw. Their findings revealed that surfactants enhanced enzymatic efficiency by lowering surface tension, which in turn facilitated better enzyme penetration into fibers, ultimately improving both pulp yield and brightness.

Further research explored the impact of surfactant type and concentration on cellulase-pretreated pulp. Agrawal et al. (2017) and Alhammad et al. (2018) demonstrated that surfactants mitigate nonspecific interactions between enzymes and lignin, thereby preventing enzyme deactivation and enhancing saccharification yields. This highlights the critical role of surfactants in optimizing enzyme accessibility and stability during biomass processing.

In a related study, Yang et al. (2019) employed *poly(dimethyl diallyl ammonium chloride)* (PDADMAC) as a process additive to support cellulase performance during pulp refining. The study revealed that PDADMAC—particularly its low-molecular-weight form—facilitated cellulase adsorption onto cellulose fibers through electrostatic attraction, effectively reducing both enzyme dosage requirements and refining energy consumption. Similarly, Wang et al. (2015) reported that the application of *cationic polyacrylamide* (CPAM) during cellulase-assisted treatment of dissolving pulp promoted enzyme adsorption via a patching or bridging mechanism. This interaction enhanced Fock reactivity while decreasing pulp viscosity, demonstrating improved fiber reactivity and process efficiency.

Moreover, Wang et al. (2015) further established that combining the biodegradable surfactant *cardanol polyoxyethylene ether* with cutinase and amylase enzymes provided an effective eco-friendly strategy for deinking mixed office wastepaper. The synergistic action of the surfactant and enzymes lowered ink surface tension and improved emulsification, resulting in efficient ink detachment and dispersion, thus advancing the sustainability of paper recycling processes.

Biodeinking

Biodeinking leverages enzymes to remove ink and coatings from recycled paper. Cellulases, xylanases, and laccases are commonly used to loosen fiber surfaces and facilitate ink release. Formulations like NS-22086 and AccelleraseRTRIO™ have proven effective across various pulp types. Engineered enzyme systems improve drainage and mechanical properties, while mixed treatments yield better results than

Table 3.2 Enzyme applications in biobleaching in recent years

Enzymes	Microorganism	Dose	T (°C)	pH	Treatment time (h)	Type of pulp/ consistency (%)	Results	References
Xylanase, laccase	*Aspergillus* sp., *Phanerochaete chrysosporium* ATCC 24725	30–40 U	35	4-6.5	14	Kraft pulp (8%w/v)	Reduction in kappa number	García-Rivero et al. (2015)
Xylanase, laccase	*Bacillus halodurans* FNP 135 and *Bacillus* sp.	15U, 20 nkat	70	9	3	Eucalyptus kraft pulp	Brightness (13%), whiteness (106.15%), breaking length (49%), burst factor (6.9%), tear factor (23%), and viscosity (11.68%); reduced kappa Kappa number	Gupta et al. (2015b)
Xylanase, Arabinofuranosidase, Acetyl xylan esterase	*Bacillus* sp. NIORKP76	5U/g	40	8	2	Kraft pulp	Kappa number reduced by 55% after multienzyme treatment, hypochlorite consumption reduced by 50%	Parab and Khandeparker (2021)

(continued)

Table 3.2 (continued)

Enzymes	Microorganism	Dose	T (°C)	pH	Treatment time (h)	Type of pulp/ consistency (%)	Results	References
Xylarase, Pectirase	*Bacillus firmus* SDB9	2.7 IU/g 15U/g	50	8.0	2	Unbleached Kraft pulp	7.9% reduction in kappa value, 2.7% increase in pulp brightness, 0.83% increase in pulp viscosity	Bhagat et al. (2021)
Xylarase, Mannanase	*Enterobacter ludwigii MY271*	10U/g	50	7.0	2h	Kraft softwood pulp (10% w/v)	Improved pulp brightness, no remarkable effect on tensile strength Brightness and whiteness increased by 11% and 75%, respectively; kappa value and chlorine usage decreased by 45.64% and 30%	Yang et al. (2017)

(continued)

Table 3.2 (continued)

Enzymes	Microorganism	Dose	T (°C)	pH	Treatment time (h)	Type of pulp/ consistency (%)	Results	References
Xylanase, Pectinase	*Xylano-pectinolytic enzymes Bacillus pumilus AJK (MTCC 10, 414)*	5.0, 1.66 IU/g	55	8.5	3 h	Wheat straw pulp	14.75% decline in kappa number; active chlorine reduced by 25%; increases in burst index by 7.98%, tear index by 3.42%, breaking length by 5.30%, and viscosity by 11.22%; decreases in BOD by 20.74% and COD by 17.87%	Sharma et al. (2022)
Xylano-pectinolytic enzymes *Bacillus pumilus*	*Bacillus amyloliquefaciens* ADI2	15, 19.5 U/g	40	8.5	3 h	Oil palm empty fruit bunches	Improved brightness by 11.25%, decreased chemical consumption by 11.25%	Nawawi et al. (2021)

(continued)

Table 3.2 (continued)

Enzymes	Microorganism	Dose	T (°C)	pH	Treatment time (h)	Type of pulp/ consistency (%)	Results	References
Laccase, Xylanase, Mannanase	*Bacillus tequilensis LXM55*	50 U/g	60	8.0	4 h	Mixed wood pulp (5% w/v)	46.32 (L + X)/ 40.25% (L + M) reduction in kappa number, 13.21/ 10.01% and 3.36/ 2.76% improvement in brightness and whiteness	Angural et al. (2020b)

(continued)

Table 3.2 (continued)

Enzymes	Microorganism	Dose	T (°C)	pH	Treatment time (h)	Type of pulp/ consistency (%)	Results	References
Laccase (L), Xylanase (X), Mannanase (M)	*Bacillus* sp. LX and *Bacillus* sp. LM	1 mL, (3:1)	70	8.5	4 h	Mixed wood pulp (5% w/v)	Brightness and whiteness increased by 11% and 4%, respectively; kappa number and chlorine consumption decreased by 49% and 40%	Angural et al. (2020a)
Xylanase, Mannanase	*Bacillus sp. NG- 27, Bacillus nealsonii PN-11*	55 IU/ g	65	8.0	1 h	Mixed wood	Brightness and whiteness increased by 11% and 75%, respectively; kappa value and chlorine usage decreased by 45.64% and 30%	Angural et al. (2021)

Yang et al. (2023). Reproduced with permission

Table 3.3 Use of chemical additives and enzymes in the papermaking process

Chemical additives	Enzymes	Process	Dose	Temperature (°C)	pH	Treatment time	Type of pulp/ consistency (%)	Results	References
Chlorine dioxide	Xylanase	Bleaching	10 IU/ g pulp	50	8.5	60 min	Hardwood kraft pulp/ 10%	20% reduction in chlorine dioxide use	Madlala et al. (2001)
Hydrogen peroxide	Laccase–mediator system	Bleaching	2.5 U/g pulp	55	6.0	90 min	Softwood kraft pulp/ 10%	Improved brightness by 6–8% ISO	Ravalason et al. (2012)
Sodium hydroxide	Pectinase	Deinking	50 U/g pulp	45	7.5	45 min	Mixed waste paper/ 12%	15% decrease in ink specks, better drainage	Li et al. (2013b)
Cationic polymer	Xylanase + Laccase	Refining	10 IU/ g pulp	50	8.0	60 min	Wheat straw pulp/10%	Enhanced tensile and burst strength	Wang et al. (2018)
Surfactant	Lipase	Deinking	25 U/g pulp	40	7.0	30 min	Mixed office waste/ 8%	Higher brightness (3 units) and fiber recovery	Nathan and Rani (2020)
Hydrogen peroxide	Cellulase-free xylanase	Biobleaching	30 IU/ g pulp	55	9.0	90 min	Bamboo kraft pulp/ 10%	Reduced kappa number by 12%, improved brightness	Dutta et al. (2020)
Sodium silicate	Xylanase + Pectinase	Bleaching	40 IU/ g pulp	50	8.0	60 min	Rice straw pulp/10%	25% decrease in chemical use, improved optical properties	Nagpal et al. (2021)

(continued)

Table 3.3 (continued)

Chemical additives	Enzymes	Process	Dose	Temperature (°C)	pH	Treatment time	Type of pulp/ consistency (%)	Results	References
Cationic starch	Mannanase	Refining	20 IU/ g pulp	45	7.0	45 min	Mixed wood pulp/8%	Increased fiber flexibility and bonding strength	Angural et al. (2021)
Hydrogen peroxide	Enzyme cocktail (Lignolytic + Hemicellulolytic)	Bleaching	15 IU/ g pulp	55	8.5	75 min	Oil palm EFB pulp/ 10%	20% improvement in brightness and strength	Nawawi et al. (2021)
Acetic acid	Feruloyl esterase	Pulp modification	5 U/g pulp	50	7.0	60 min	Dissolving pulp/8%	Improved fiber flexibility and reactivity	Hou et al. (2022)

Yang et al. (2023). Reproduced with permission

single enzymes. LPMO integration enhances fiber uniformity and supports nanofibrillation (Wang et al. 2017; Buzała et al. 2015, 2016; Taneda et al. 2012; Hu et al. 2018).

Sticky Control

Stickies—adhesive contaminants from recycled pulp—pose serious operational challenges. They originate from resins, sizing agents, and synthetic compounds like PSAs and hot melts. Upon destabilization, stickies deposit on equipment, impairing pulp quality and increasing treatment costs. Enzymes offer an eco-friendly solution by hydrolyzing ester bonds in sticky materials (Jidong et al. 2011; Zhang et al. 2017; Hamann and Strauss 2003).

Lipolytic enzymes and pectinases are central to sticky control. Commercial products such as OptimyzeR, StickAwayR, and EnzynkR contain esterases, lipases, cellulases, hemicellulases, amylases, and oxidoreductases. Lab-scale treatments with lipase and esterase removed up to 90% of stickies and reduced PSA adhesion. Combined enzyme systems also reduced microsticky size and stabilized white water dynamics. A blend of thermophilic esterases with amylase, pectinase, and xylanase achieved 76.5% sticky degradation and improved pulp properties (Tang et al. 2019; Ebrahimi et al. 2015; Xiang et al. 2006; Zhang et al. 2017).

Three comparative tables are presented below to illustrate the current state of enzyme applications in the pulp and paper industry. Table 3.4 highlights the technology readiness levels of enzymes in pulp and paper manufacturing, showing which biocatalysts are already deployed at mill scale, which remain at pilot stage, and which are still confined to laboratory research. Table 3.5 provides an overview of global manufacturers and their commercial enzyme portfolios for the pulp and paper sector, mapping out the major suppliers and their product lines across bleaching, deinking, refining, and starch modification. Finally, Table 3.6 integrates these perspectives by presenting the industrial deployment status of enzymes alongside the leading market players in papermaking applications, offering a consolidated view of readiness and supplier strength across the industry.

3.3   Factors Influencing Enzyme Cocktail Performance

Effective application of enzyme cocktails in pulp and paper (P&P) processing depends on multiple interrelated factors. Enzyme synergy is influenced by type, concentration, sequence of addition, and dosage ratios. For example, laccase mediator systems (LMS) may inactivate xylanase or even laccase itself, depending on the order of application, affecting delignification, brightness, and kappa number (Bajpai et al. 2006; Woolridge 2014; Immerzeel and Fiskari 2023). Optimizing enzyme combinations requires precise data and consideration of environmental conditions, pulp characteristics, and chemical additives (Wei et al. 2021).

Table 3.4 Readiness levels of enzyme technologies in pulp and paper

Enzyme	Substrate	Functions	Applications in pulp and paper industry	Status (Mill/Pilot/Lab)
Cellulase	Cellulose	Partial hydrolysis of cellulose	Ink release from fiber surface; hydrolysis of colloidal material in mill drainage	Mill
Xylanase	Xylan	Degradation of redeposited xylan & lignin–carbohydrate complexes	Refining/fiber modification; vessel picking; deinking; drainage improvement; bleach boosting; dissolving pulp production; biopulping; shives removal; debarking	Mill/Pilot/Lab
Mannanase	Glucomannan	Removal of glucomannan	Bleach boosting; biopulping	Mill/Pilot
Laccase	Lignin	Degradation of lignin (mediator-assisted)	Bleaching; effluent treatment	Pilot
Mn-peroxidase	Lignin	Degradation of lignin (additive-assisted)	Bleaching; effluent treatment	Lab
Lipase	Fat/oil	Hydrolysis of triglycerides	Pitch control; contaminant control; deinking of oil-based ink	Mill
Amylase	Starch	Hydrolysis of α-1,4/α-1,6 bonds	Surface sizing; starch coating; deinking; drainage improvement; slime control	Mill
Esterase	Macrostickies	Breaks ester bonds	Stickies control	Mill
Protease	Protein	Hydrolysis of cell wall proteins	Biofilm removal	Mill
Pectinase	Pectin	Hydrolysis of cambial layer	Refining; debarking; energy saving; reduced cationic demand	Mill/Lab

pH, Temperature, and Reaction Time

Enzyme activity is highly sensitive to pH, temperature, and residence time. Laccase exhibits optimal activity within a pH range of 2–5 and at temperatures between 50–80 °C, whereas xylanase demonstrates system-dependent variation, functioning at pH 6–9 in prokaryotes and pH 4–6 in eukaryotes, with an effective temperature range of 35–60 °C. Process-specific conditions—such as alkaline pulping or ambient-temperature

Table 3.5 Global enzyme manufacturers serving the pulp and paper market

Company and headquarters	Enzymes	Trade names/applications
Enzymatic Deinking Technologies (USA)	Cellulases, Hemicellulases, Lipases, Laccases	FibreZyme® G4 (Neutral cellulases for deinking)
Dyadic International (USA) FibreZyme® G200 (Neutral cellulases for refining & paper strength)	Cellulases, Xylanases, Amylases, Proteases	Novozymes Pulpzyme® (Xylanases for bleaching), Novozymes Resinase® (Lipases for pitch control)
DuPont (Genencor, USA)	Cellulases, Proteases, Xylanases	ArrowPulp (Bleaching), ArrowDeink (Deinking), ArrowRefine (Refining), ArrowSize (Starch cooking)
EpygenLabs (UAE)	Cellulases, Hemicellulases, Xylanases, Esterases	Papyrase® series (Fiber modification, bleaching, stickies control, starch conversion)
Metgen Oy (Finland)	Laccases, Oxidoreductases	MetZyme® series (Bleaching, deinking, starch conversion, peroxide removal) MetZyme® LIGNO™, BRILA™, POVON™
AB Enzymes (Germany)	Amylases, Xylanases, Proteases, Lipases, Cellulases, Pectinases	Ecopulp (Bleaching)
Advanced Enzyme Technologies (India)	Xylanases, Cellulases, Amylase, Pectinases, Lipases	SEBrite series (Bleach boosting, deinking, starch modification, stickies removal)
ANC Enzymes (Singapore)	Amylases, Xylanase, Laccase, Proteases	SEBrite STR (Lipases for stickies removal)
Anil Bioplus (India)	Amylases, Cellulases, Pectinase, Proteases	Pulpase DI/RF/BL (Deinking, refining, bleaching)
Leveking Enzymes (China)	Lipases, Xylanases, Amylases	LPK-CD05 (Deinking), LPK-CR01 (Refining), LPK-CS01S (Surface sizing)
Rossari Biotech (India)	Xylanases, Cellulases, Proteases	Rossapase series (Deinking, bleaching, refining)
Specialty Enzymes (USA)	Xylanases, Cellulases, Amylase, Pectinases	SEBrite Bleach (Bleaching), SEBrite DI (Deinking)
Sukahan Biotech (China)	Xylanases, Proteases, Cellulases	SUKAZYM series (Bleaching, deinking, starch conversion)
Tex Biosciences (India)	Proteases, Lipases, Xylanases	Texzyme series (Deinking, bleaching, refining, starch modification, stickies control)

Table 3.6 Industrial Deployment Status of Enzymes in Papermaking and Leading Market Players

Enzyme type	Primary application in papermaking	Industrial deployment status	Leading market players
Cellulases	Fiber modification, drainage improvement, refining energy reduction	Widely deployed in kraft and tissue paper production	Novozymes, AB Enzymes, BASF, Dyadic International
Xylanases	Bleaching, brightness enhancement, chemical savings	Mature deployment in bleaching sequences (ECF/TCF processes)	Novozymes, Enzymatic Deinking Technologies, Rajiv Gandhi Biotech Centre collaborations
Laccases	Biobleaching, pitch control, lignin modification	Emerging deployment; pilot-to-commercial scale in specialty papers	Novozymes, MetGen, Dyadic International
Pectinases	Deinking, stickies removal, fiber recovery	Commercial deployment in newsprint and recycled paper	AB Enzymes, Buckman, Enzymatic Deinking Technologies
Amylases	Starch modification, coating improvement	Established deployment in coating and surface sizing	BASF, Novozymes, Specialty Enzymes & Biotechnologies
Lipases	Pitch and stickies control	Limited but growing deployment in mills with high resin content	Buckman, MetGen, Enzymatic Deinking Technologies
Proteases	Fiber surface modification, contaminant removal	Niche deployment; under evaluation for broader applications	Dyadic International, Specialty Enzymes & Biotechnologies

papermaking—require tailored enzyme formulations. Reaction time must align with mill operations; for example, xylanase bleaching typically requires 2–3 hours, but co-application with cellulase can reduce this by 1 hour (Bajpai 1999; Tolan and Guenette 1997; Nagar et al. 2011).

Pulp Type

Enzyme efficacy varies with pulp type, fiber composition, and crystallinity. Wood and nonwood pulps (e.g., pine, bamboo, bagasse) differ in hemicellulose and lignin content, affecting enzyme accessibility. Studies show softwood sulfite pulp responds better to enzymatic refining than sulfate pulp. EG and xylanase treatments reduced viscosity and improved reactivity in eucalyptus and sisal dissolving pulps. Xylanase pretreatment was most effective on bagasse pulp, while LMS degraded lipophilic compounds across various pulps. Wastepaper biodeinking efficiency also varies: laser-printed paper showed 86.6% efficiency, while newspaper achieved only 12.9% (Gao et al. 2018; Lee et al. 2013; Kenney et al. 2013; Haske-Cornelius et al. 2020; Ibarra et al. 2009; Madlala et al. 2001; Gutierrez et al. 2006).

Chemical Additives

Chemical agents used in pulping, bleaching, and deinking—such as NaOH, AQ, ClO_2, H_2O_2, and surfactants—can inhibit enzyme activity through interactions with phenols, heavy metals, and aromatic compounds. Conversely, enzymes can modulate additive performance. Pectinase improves retention aid efficiency and reduces stickies. Surfactants enhance enzymatic action by lowering surface tension and nonspecific binding, improving pulp yield and saccharification. Additives like PDADMAC and CPAM increase cellulase adsorption, reducing enzyme dosage and energy use. Cardanol-based surfactants combined with cutinase and amylase improved deinking beyond chemical methods (Kumar and Rani 2019; Wang et al. 2018; Ricard et al. 2005; Nagpal et al. 2021; Agrawal et al. 2017).

Auxiliary Technologies

Integrating enzymes with auxiliary technologies—such as mechanical refining, fiber fractionation, ultrasonic treatment, and chemical pretreatment—can significantly enhance pulp and paper processing efficiency. Studies show that pre-deconstruction of wood chips before xylanase application improves pulp brightness and strength while reducing operational costs. Depth refining enhances cellulase accessibility by exposing hydroxyl groups, leading to superior paper properties. Sequential treatments combining acetic acid, PFI milling, and enzymes yield higher fermentable sugar output. Similarly, ozone combined with cellulases improves fiber flexibility in bamboo kraft pulp (Pandey et al. 2022; An et al. 2022; Wang et al. 2020; Su et al. 2022).

3.4 Advancements in Biotechnological Practices in the Pulp and Paper Field

Natural enzymes often lack the stability and specificity required for harsh industrial conditions in P&P. Advanced enzyme engineering—such as directed evolution, rational design, and machine learning—has enabled the development of robust, thermotolerant, and pH-resistant enzymes tailored for industrial use (Palackal et al. 2004; Sharma et al. 2021). AI-driven computational design is emerging as a transformative tool for optimizing enzyme performance and reducing development time.

Modifications to enzyme structure—such as widening active sites, enhancing electrostatic interactions, and introducing stabilizing bonds—can improve substrate specificity and catalytic efficiency Domain swapping, as demonstrated in fungal laccases, has extended enzyme half-life and broadened pH tolerance for kraft delignification (Zhu et al. 2022; Pardo et al. 2018).

High-yield enzyme expression is achievable through synthetic biology, CRISPR/Cas optimization, and fermentation parameter tuning. Multienzyme assemblies and enzyme recycling strategies—such as recovering xylanase from filtrates—offer cost-effective solutions for continuous pulp and paper operations (Dixit et al. 2022; Yang et al. 2019; Gomes et al. 2015).

3.5 Future Considerations

As the global focus shifts toward energy efficiency and carbon reduction, the pulp and paper industry is embracing enzymatic technologies for cleaner, low-emission manufacturing. Despite decades of enzyme use, optimizing enzyme cocktails remains a site-specific challenge requiring tailored solutions for each mill (Immerzeel and Fiskari 2023).

Future progress hinges on comprehensive life cycle assessments and integration into circular economy frameworks. Leveraging AI and machine learning to model enzyme interactions in complex pulp and paper environments will accelerate innovation, improve process efficiency, and reduce research costs—paving the way for sustainable, data-driven bioprocessing.

References

Abdulhadi Y, Ashish V (2021) Production, optimization and deinking capacity of alkaline cellulase produced from *Mucor circinelloides* WSSDBS2F1. Cellul Chem Technol 55:605–618. https://doi.org/10.35812/CelluloseChemTechnol.2021.55.49

Agrawal R, Satlewal A, Kapoor M, Mondal S, Basu B (2017) Investigating the enzyme-lignin binding with surfactants for improved saccharification of pilot scale pretreated wheat straw. Bioresour Technol 224:411–418. https://doi.org/10.1016/j.biortech.2016.11.026

Ahlawat S, Mandhan RP, Dhiman SS, Kumar R, Sharma J (2008) Potential application of alkaline pectinase from *Bacillus subtilis* SS in pulp and paper industry. Appl Biochem Biotechnol 149:287–293. https://doi.org/10.1007/s12010-007-8096-9

Akbarpour I, Resalati H (2022) The effect of enzymatic pre-treatment with amylase and refining on the physical and dewatering properties of OCC pulp. Iran J Wood Pap Ind 12:507–519

Alhammad A, Adewale P, Kuttiraja M, Christopher LP (2018) Enhancing enzyme-aided production of fermentable sugars from poplar pulp in the presence of non-ionic surfactants. Bioproc Biosyst Eng 41:1133–1142. https://doi.org/10.1007/s00449-018-1942-z

Ali S, Khatri Z, Khatri A, Tanwari A (2014) Integrated desizing–bleaching–reactive dyeing process for cotton towel using glucose oxidase enzyme. J Clean Prod 66:562–567. https://doi.org/10.1016/j.jclepro.2013.11.035

An X, Zhang R, Liu L, Yang J, Tian Z, Yang G, Cao H, Cheng Z, Ni Y, Liu H (2022) Ozone pretreatment facilitating cellulase hydrolysis of unbleached bamboo pulp for improved fiber flexibility. Ind Crop Prod 178:114577. https://doi.org/10.1016/j.indcrop.2022.114577

Angural S, Kumar A, Kumar D, Warmoota R, Sondhi S, Gupta N (2020a) Lignolytic and hemicellulolytic enzyme cocktail production from *Bacillus tequilensis* LXM 55 and its application in pulp biobleaching. Bioproc Biosyst Eng 43:2219–2229. https://doi.org/10.1007/s00449-020-02407-4

Angural S, Rana M, Sharma A, Warmoota R, Puri N, Gupta N (2020b) Combinatorial biobleaching of mixedwood pulp with lignolytic and hemicellulolytic enzymes for paper making. Indian J Microbiol 60:383–387. https://doi.org/10.1007/s12088-020-00867-6

Angural S, Bala I, Kumar A, Kumar D, Jassal S, Gupta N (2021) Bleach enhancement of mixed wood pulp by mixture of thermo-alkali-stable xylanase and mannanase derived through co-culturing of Alkalophilic *Bacillus* sp. NG-27 and *Bacillus nealsonii* PN-11. Heliyon 7:e05673

Bajpai P (1999) Application of enzymes in the pulp and paper industry. Biotechnol Prog 15:147–157. https://doi.org/10.1021/bp990013k

Bajpai PK (2011) Emerging applications of enzymes for energy saving in pulp & paper industry. IPPTA 23:181–186

Bajpai P (2018) Biotechnology for pulp and paper processing. Springer

Bajpai P, Anand A, Bajpai PK (2006) Bleaching with lignin-oxidizing enzymes. Biotechnol Annu Rev 12:349–378. https://doi.org/10.1016/S1387-2656(06)12010-4

Ballinas-Casarrubias L, Gonzalez-Sanchez G, Eguiarte-Franco S, Siqueiros-Cendon T, Flores-Gallardo S, Villa ED, Hernandez MDD, Rocha-Gutierrez B, Rascon-Cruz Q (2020) Chemical characterization and enzymatic control of stickies in kraft paper production. Polymers 12:245. https://doi.org/10.3390/polym12010245

Bhagat DD, Dudhagara PR, Desai PV (2021) Statistical approach for pectinase production by *Bacillus firmus* SDB9 and evaluation of pectino-xylanolytic enzymes for pretreatment of kraft pulp. J Microbiol Biotechnol Food Sci 2021:396–406

Infinita Biotech (2023) Applications of enzymes in paper and pulp industries. https://infinitabiotech.com/blog/applications-of-enzymes-in-paper-and-pulp-industries/

Buzała K, Przybysz P, Rosicka-Kaczmarek J, Kalinowska H (2015) Production of glucose-rich enzymatic hydrolysates from cellulosic pulps. Cellulose 22:663–674

Buzała KP, Przybysz P, Kalinowska H, Derkowska M (2016) Effect of cellulases and xylanases on refining process and kraft pulp properties. PLoS ONE 11:e0161575. https://doi.org/10.1371/journal.pone.0161575

Buzała KP, Kalinowska H, Przybysz P, Małachowska E (2017) Conversion of various types of lignocellulosic biomass to fermentable sugars using kraft pulping and enzymatic hydrolysis. Wood Sci Technol 51:873–885. https://doi.org/10.1007/s00226-017-0916-7

Cebreiros F, Seiler S, Dalli SS, Lareo C, Saddler J (2021) Enhancing cellulose nanofibrillation of eucalyptus Kraft pulp by combining enzymatic and mechanical pretreatments. Cellulose 28:189–206

Cedillo VB, Prieto A, Martinez MJ (2013) Potential of *Ophiostoma piceae* sterol esterase for biotechnologically relevant hydrolysis reactions. Bioengineered 4:249–253. https://doi.org/10.4161/bioe.22818

Chauhan PS, Goradia B, Saxena A (2017) Bacterial laccase: recent update on production, properties and industrial applications. 3 Biotech 7(5):323. https://doi.org/10.1007/s13205-017-0955-7

Chen Y, Wan J, Ma Y, Tang B, Han W, Ragauskas AJ (2012) Modification of old corrugated container pulp with laccase and laccase–mediator system. Bioresour Technol 110:297–301. https://doi.org/10.1016/j.biortech.2011.12.120

Chen YA, Zhou Y, Qin Y, Liu D, Zhao X (2018) Evaluation of the action of Tween 20 non-ionic surfactant during enzymatic hydrolysis of lignocellulose: pretreatment, hydrolysis conditions and lignin structure. Bioresour Technol 269:329–338. https://doi.org/10.1016/j.biortech.2018.08.119

Chen G, Dong J, Wan J, Ma Y, Wang Y (2021) Fiber characterization of old corrugated container bleached pulp with laccase and glycine pretreatment. Biomass Convers Biorefinery 13:583–592

Chen H, Li S, Cui Z, Feng T, Wang H, Ni Z, Gao E, Fang Z (2022) Synergistic degradation of maize straw lignin by manganese peroxidase from *Irpex lacteus*. Appl Biochem Biotechnol 195:3855–3871. https://doi.org/10.1007/s12010-022-04189-9

Chen YL, Chen JC, Pang ZQ, Yang GH (2014) Effects of cellulase treatment on surface structure of fast-growing poplar APMP pulp. In: Advanced materials research. Trans Tech Publications, Zurich, pp 533–536

Clark T, Steward D, Bruce M, McDonald A, Singh A, Senior D (1991) Improved bleachability of radiata pine kraft pulps following treatment with hemicellulolytic enzymes. Appita J 44:389–403

Collins T, Gerday C, Feller G (2005) Xylanases, xylanase families and extremophilic xylanases. FEMS Microbiol Rev 29:3–23. https://doi.org/10.1016/j.femsre.2004.06.005

Colonia BSO, Woiciechowski AL, Malanski R, Letti LAJ, Soccol CR (2019) Pulp improvement of oil palm empty fruit bunches associated to solid-state biopulping and biobleaching with xylanase and lignin peroxidase cocktail produced by *Aspergillus* sp. LPB-5. Bioresour Technol 285:121361. https://doi.org/10.1016/j.biortech.2019.121361

Cui L, Meddeb-Mouelhi F, Beauregard M (2016) Permutation of refining and cellulase treatments determines the overall impact on drainability and strength properties in kraft pulp. Nord Pulp Pap Res J 31:315–322. https://doi.org/10.3183/npprj-2016-31-02-p315-322

Creative Enzymes (2023) Application of enzymes in pulp and paper industry. https://www.creative-enzymes.com/resource/application-of-enzymes-in-pulp-and-paper-industry_64.html

Dashtban M, Schraft H, Syed TA, Qin W (2010) Fungal biodegradation and enzymatic modification of lignin. Int J Biochem Mol Biol 1:36–50

Dixit M, Gupta GK, Yadav M, Chhabra D, Kapoor RK, Pathak P, Bhardwaj NK, Shukla P (2022) Improved deinking and biobleaching efficiency of enzyme consortium from *Thermomyces lanuginosus* VAPS25 using genetic algorithm-artificial neural network based tools. Bioresour Technol 349:126846. https://doi.org/10.1016/j.biortech.2022.126846

Dutta PD, Neog B, Goswami T (2020) Xylanase enzyme production from *Bacillus australimaris* P5 for prebleaching of bamboo (Bambusa tulda) pulp. Mater Chem Phys 243:122227

Ebrahimi M, Ramezani O, Rahmaninia M, Kermanian H, Andalibian MA (2014) Performance of amylase on properties of recycled OCC pulp pre-soaked at different pH(s). For Wood Prod 67(2):325–333. https://doi.org/10.22059/jfwp.2014.51549

Ebrahimi BR, Resalati H, Aryaie MM, Ghasemiyan A, Shakeri A (2015) Enzymatic control of stickies in recycle paper. Iran J Chem Eng 14(78):4–12. https://www.sid.ir/paper/16246/en

Fang J, Huang F, Gao P (1999) Optimization of cellobiose dehydrogenase production by *Schizophyllum commune* and effect of the enzyme on kraft pulp bleaching by ligninases. Process Biochem 34:957–961. https://doi.org/10.1016/S0032-9592(99)00016-3

Flitsch A, Prasetyo EN, Sygmund C, Ludwig R, Nyanhongo GS, Guebitz GM (2013) Cellulose oxidation and bleaching processes based on recombinant *Myriococcum thermophilum* cellobiose dehydrogenase. Enzym Microb Technol 52:60–67. https://doi.org/10.1016/j.enzmictec.2012.10.007

Gao S, Li L, Lin Y, Han S (2018) Deinkability of different secondary fibers by enzymes. Nord Pulp Pap Res J 33:12–20. https://doi.org/10.1515/npprj-2018-3014

García-Rivero M, Venegas IM, Carmona SV, Jiménez GZ, Segura PZ, Trujillo MM (2015) Enzymatic pretreatment to enhance chemical bleaching of a kraft pulp. Rev Mex Ing Quim 14:335–345

Gomes D, Rodrigues AC, Domingues L, Gama M (2015) Cellulase recycling in biorefineries—is it possible? Appl Microbiol Biotechnol 99:4131–4143. https://doi.org/10.1007/s00253-015-6535-z

Gricajeva A, Nadda AK, Gudiukaite R (2022) Insights into polyester plastic biodegradation by carboxyl ester hydrolases. J Chem Technol Biotechnol 97:359–380. https://doi.org/10.1002/jctb.6745

Guo ZP, Duquesne S, Bozonnet S, Nicaud JM, Marty A, O'Donohue MJ (2017) Expressing accessory proteins in cellulolytic *Yarrowia lipolytica* to improve the conversion yield of recalcitrant cellulose. Biotechnol Biofuels 10:298. https://doi.org/10.1186/s13068-017-0990-y

Gupta C, Jain P, Kumar D, Dixit A, Jain R (2015a) Production of cellulase enzyme from isolated fungus and its application as efficient refining aid for production of security paper. Int J Appl Microbiol Biotechnol Res 3:11–19

Gupta V, Garg S, Capalash N, Gupta N, Sharma P (2015b) Production of thermo-alkali-stable laccase and xylanase by co-culturing of *Bacillus* sp. and *B. halodurans* for biobleaching of kraft pulp and deinking of waste paper. Bioproc Biosyst Eng 38:947–956. https://doi.org/10.1007/s00449-014-1340-0

Gupta A, Tiwari A, Ghosh P, Arora K, Sharma S (2023) Enhanced lignin degradation of paddy straw and pine needle biomass by combinatorial approach of chemical treatment and fungal enzymes for pulp making. Bioresour Technol 368:128314. https://doi.org/10.1016/j.biortech.2022.128314

Gutiérrez A, del Río JC, Rencoret J, Ibarra D, Martínez A (2006) Main lipophilic extractives in different paper pulp types can be removed using the laccase–mediator system. Appl Microbiol Biotechnol 72:845–851. https://doi.org/10.1007/s00253-006-0346-1

Hamann L, Strauss J (2003) Stickies: definitions, causes and control options. Wochenbl Pap 131:652–663

Haske-Cornelius O, Hartmann A, Brunner F, Pellis A, Bauer W, Nyanhongo GS, Guebitz GM (2020) Effects of enzymes on the refining of different pulps. J Biotechnol 320:1–10. https://doi.org/10.1016/j.jbiotec.2020.06.006

Hou YH, Yang ZH, Wang JZ, Yang QZ (2022) Characterization of a thermostable alkaline feruloyl esterase from *Alternaria alternata* and its synergism in dissolving pulp production. Biochem Eng J 187:108657. https://doi.org/10.1016/j.bej.2022.108657

Hu J, Tian D, Renneckar S, Saddler JN (2018) Enzyme mediated nanofibrillation of cellulose by the synergistic actions of an endoglucanase, lytic polysaccharide monooxygenase (LPMO) and xylanase. Sci Rep 8:3195. https://doi.org/10.1038/s41598-018-21016-6

Hugouvieux-Cotte-Pattat N, Condemine G, Shevchik VE (2014) Bacterial pectate lyases, structural and functional diversity. Environ Microbiol Rep 6:427–440. https://doi.org/10.1111/1758-2229.12166

Ibarra D, Köpcke V, Ek M (2009) Exploring enzymatic treatments for the production of dissolving grade pulp from different wood and non-wood paper grade pulps. 10th EWLP. Stockholm, Sweden 63:721–730. https://doi.org/10.1515/HF.2009.102

Immerzeel P, Fiskari J (2023) Synergism of enzymes in chemical pulp bleaching from an industrial point of view—a critical review. Can J Chem Eng 101:312–321. https://doi.org/10.1002/cjce.24374

Janeček Š, Svensson B, MacGregor EA (2014) α-Amylase: an enzyme specificity found in various families of glycoside hydrolases. Cell Mol Life Sci 71:1149–1170. https://doi.org/10.1007/s00018-013-1388-z

Jha H, Patil M (2013) Biopulping of sugarcane bagasse and decolorization of kraft liquor by the laccase produced by *Klebsiella aerogenes* NCIM 2098. Malays J Microbiol 9:301–307. https://doi.org/10.21161/mjm.50312

Jiang J, Li Z, Fu Y, Wang Z, Qin M (2018) Enhancement of colloidal particle and lignin removal from pre-hydrolysis liquor of aspen by a combination of pectinase and cationic polymer treatment. Sep Purif Technol 199:78–83. https://doi.org/10.1016/j.seppur.2018.01.053

Jidong L, Yanling H, Jinwei Z, Wenjing D (2011) Accumulation of dissolved and colloidal substances in water recycled during papermaking. Chem Eng J 168:604–609. https://doi.org/10.1016/j.cej.2011.01.030

Kamimura N, Sakamoto S, Mitsuda N, Masai E, Kajita S (2019) Advances in microbial lignin degradation and its applications. Curr Opin Biotechnol 56:179–186. https://doi.org/10.1016/j.copbio.2018.11.011

Kenney KL, Smith WA, Gresham GL, Westover TL (2013) Understanding biomass feedstock variability. Biofuels 4:111–127

Khambhaty Y, Akshaya R, Suganya CR, Sreeram KJ, Rao JR (2018) A logical and sustainable approach towards bamboo pulp bleaching using xylanase from *Aspergillus nidulans*. Int J Biol Macromol 118:452–459. https://doi.org/10.1016/j.ijbiomac.2018.06.100

Kuhad RC, Gupta R, Singh A (2011) Microbial cellulases and their industrial applications. Enzym Res 2011:280696. https://doi.org/10.4061/2011/280696

Kumar H, Christopher LP (2017) Recent trends and developments in dissolving pulp production and application. Cellulose 24:2347–2365

Kumar A, Gautam A, Dutt D (2020) Bio-pulping: an energy saving and environment-friendly approach. Phys Sci Rev 5:1–9. https://doi.org/10.1515/psr-2019-0043

Kumar A, Ram C, Tazeb A (2021a) Enzyme-assisted pulp refining: an energy saving approach. Phys Sci Rev 6:20190046. https://doi.org/10.1515/psr-2019-0046

Kumar V, Pathak P, Harsh NSK, Bhardwaj NK (2021b) Biodeinking: an eco-friendly alternative for chemicals-based recycled fiber processing. Phys Sci Rev 2021:20190045. https://doi.org/10.1515/psr-2019-0045

Kumar NV, Rani ME (2019) Microbial enzymes in paper and pulp industries for bioleaching application. In: Research trends of microbiology. MedDocs eBooks, pp 1–11

Lee CK, Ibrahim D, Omar IC (2013) Enzymatic deinking of various types of waste paper: efficiency and characteristics. Process Biochem 48:299–305. https://doi.org/10.1016/j.procbio.2012.12.015

Lee S, Park H, Im W, Park H, Lee HL, Youn HJ (2020) Effects of enzyme mixture and beating treatment on the properties of pulp fibers. Palpu Chongi Gisul/. J Korea Tech Assoc Pulp Pap Ind 52:101–109. https://doi.org/10.7584/JKTAPPI.2020.10.52.5.101

Li H, Fu S, Peng L (2013a) Fiber modification of unbleached kraft pulp with laccase in the presence of ferulic acid. BioResources 8:5794–5806

Li Z, Liu C, Qin M, Fu Y, Gao Y (2013b) Stickies control with pectinase for improving behavior of cationic polymers in a mixture of chemithermomechanical pulp and deinked pulp. BioResources 8:189–200

Liu N, Qin M, Gao Y, Li Z, Fu Y, Xu Q (2012) Pulp properties and fiber characteristics of xylanase-treated aspen APMP. BioResources 7:3367–3377

Liu M, Yang S, Long L, Cao Y, Ding S (2018) Engineering a chimeric lipase-cutinase (Lip-Cut) for efficient enzymatic deinking of waste paper. BioResources 13:981–996

Madlala AM, Bissoon S, Singh S, Christov L (2001) Xylanase-induced reduction of chlorine dioxide consumption during elemental chlorine-free bleaching of different pulp types. Biotechnol Lett 23:345–351. https://doi.org/10.1023/A:1005693205016

Nagar S, Gupta VK (2021) Hyper production and eco-friendly bleaching of kraft pulp by xylanase from *Bacillus pumilus* SV-205 using agro waste material. Waste Biomass Valorization 12:4019–4031. https://doi.org/10.1007/s12649-020-01258-0

Nagar S, Mittal A, Kumar D, Kumar L, Kuhad RC, Gupta VK (2011) Hyper production of alkali stable xylanase in lesser duration by *Bacillus pumilus* SV-85S using wheat bran under solid state fermentation. New Biotechnol 28(5):581–587. https://doi.org/10.1016/j.nbt.2010.12.004

Nagpal R, Bhardwaj NK, Mishra OP, Mahajan R (2021) Cleaner bio-pulping approach for the production of better strength rice straw paper. J Clean Prod 318:128539. https://doi.org/10.1016/j.jclepro.2021.128539

Nathan VK, Rani ME (2020) A cleaner process of deinking waste paper pulp using *Pseudomonas mendocina* ED9 lipase supplemented enzyme cocktail. Environ Sci Pollut Res 27:36498–36509. https://doi.org/10.1007/s11356-020-09641-z

Nawawi MH, Mohamad R, Tahir PM, Asa'ari AZ, Saad WZ (2021) Pulp enhancement of oil palm empty fruit bunches (OPEFBs) via biobleaching by using xylano-pectinolytic enzymes of *Bacillus amyloliquefaciens* ADI2. Molecules 26(14):4279. https://doi.org/10.3390/molecules26144279

Paice M, Jurasek L (1984) Removing hemicellulose from pulps by specific enzymic hydrolysis. J Wood Chem Technol 4(2):187–198. https://doi.org/10.1080/02773818408081152

Palackal N, Brennan Y, Callen WN, Dupree P, Frey G, Goubet F, Hazlewood GP, Healey S, Kang YE, Kretz KA, Lee E, Tan X, Tomlinson GL, Verruto J, Wong VW, Mathur EJ, Short JM, Robertson DE, Steer BA (2004) An evolutionary route to xylanase process fitness. Protein Sci 13(2):494–503. https://doi.org/10.1110/ps.03333504

Pandey LK, Kumar A, Dutt D, Singh SP (2022) Influence of mechanical operation on the biodelignification of *Leucaena leucocephala* by xylanase treatment. 3 Biotech 12:20. https://doi.org/10.1007/s13205-021-03024-y

Parab P, Khandeparker R (2021) Xylanolytic enzyme consortia from *Bacillus* sp. NIORKP76 for improved biobleaching of kraft pulp. Bioproc Biosyst Eng 44:2513–2524. https://doi.org/10.1007/s00449-021-02623-6

Pardo I, Rodríguez-Escribano D, Aza P, De Salas F, Martínez AT, Camarero S (2018) A highly stable laccase obtained by swapping the second cupredoxin domain. Sci Rep 8(1):1–10. https://doi.org/10.1038/s41598-018-34008-

Peng YY, Liu H, Fu SY, Li XL (2018) Effect of enzyme-assisted refining on the properties of bleached softwood pulp. Paper Biomater 3(1):7–15. https://doi.org/10.26599/PBM.2018.9260016

Ravalason H, Herpoel-Gimbert I, Record E, Bertaud F, Grisel S, de Weert S, van den Hondel CA, Asther M, Petit-Conil M, Sigoillot JC (2009) Fusion of a family 1 carbohydrate binding module of *Aspergillus niger* to the *Pycnoporus cinnabarinus* laccase for efficient softwood kraft pulp biobleaching. J Biotechnol 142:220–226. https://doi.org/10.1016/j.jbiotec.2009.04.013

Ravalason H, Bertaud F, Herpoel-Gimbert I, Meyer V, Ruel K, Joseleau JP, Grisel S, Olive C, Sigoillot JC, Petit-Conil M (2012) Laccase/HBT and laccase-CBM/HBT treatment of softwood kraft pulp: impact on pulp bleachability and physical properties. Bioresour Technol 121:68–75. https://doi.org/10.1016/j.biortech.2012.06.077

Ribeiro RSA, Pohlmann BC, Calado V, Bojorge N, Pereira N Jr (2019) Production of nanocellulose by enzymatic hydrolysis: trends and challenges. Eng Life Sci 19(4):279–291. https://doi.org/10.1002/elsc.201800158

Ricard M, Orccotoma J, Ling J, Watson R (2005) Pectinase reduces cationic chemical costs in peroxide-bleached mechanical grades. Pulp Paper Canada 106(12):T264-268

Rodrigues IDSV, Silva CGS, da Silva RS, Dolabella SS, Fernandes MF, Fernandes RPM (2019) Screening of bacterial extracellular xylanase producers with potential for cellulose pulp biobleaching. Acta Scientiarum. Biol Sci 41:e42101. https://doi.org/10.4025/actascibiolsci.v41i1.42101

Sahinkaya M, Colak DN, Ozer A, Canakci S, Deniz I, Belduz AO (2019) Cloning, characterization and paper pulp applications of a newly isolated DyP type peroxidase from *Rhodococcus* sp. T1. Mol Biol Rep 46:569–580. https://doi.org/10.1007/s11033-018-4509-

Satapathy S, Rout JR, Kerry RG, Thatoi H, Sahoo SL (2020) Biochemical prospects of various microbial pectinase and pectin: an approachable concept in pharmaceutical bioprocessing. Front Nutr 7:117. https://doi.org/10.3389/fnut.2020.00117

Schou C, Schülein M, Vollmond T (1994) A novel cellobiose oxidase, an enzymatic agent and a process for treating paper pulp (U.S. Patent No. 5,356,800). Google Patents. https://patents.google.com/patent/US5356800A/en

Sharma A, Gupta G, Ahmad T, Mansoor S, Kaur B (2021) Enzyme engineering: current trends and future perspectives. Food Rev Intl 37(2):121–154. https://doi.org/10.1080/87559129.2019.1695835

Sharma D, Nagpal R, Agrawal S, Bhardwaj N, Mahajan R (2022) Eco-friendly bleaching of agrowaste wheat straw using crude alkalo-thermotolerant cellulase-free xylano-pectinolytic enzymes. Appl Biochem Biotechnol 194:620–634. https://doi.org/10.1007/s12010-021-03641-6

Singh G, Arya SK (2019) Utility of laccase in pulp and paper industry: a progressive step towards the green technology. Int J Biol Macromol 134:1070–1084. https://doi.org/10.1016/j.ijbiomac.2019.05.168

Singh G, Kaur S, Khatri M, Arya SK (2019) Biobleaching for pulp and paper industry in India: emerging enzyme technology. Biocatal Agric Biotechnol 17:558–565. https://doi.org/10.1016/j.bcab.2019.01.019

Singh G, Capalash N, Kaur K, Puri S, Sharma P (2016a) Enzymes: applications in pulp and paper industry. In: Agro-industrial wastes as feedstock for enzyme production. Elsevier, pp 157–172

Singh S, Singh VK, Aamir M, Dubey Mk, Patel JS, Upadhyay RS, Gupta VK (2016b) Cellulase in pulp and paper industry. In: New and future developments in microbial biotechnology and bioengineering. Elsevier, pp 153–163

Smit AT, Huijgen WJJ (2017) The promotional effect of water-soluble extractives on the enzymatic cellulose hydrolysis of pretreated wheat straw. Biores Technol 243:994–999. https://doi.org/10.1016/j.biortech.2017.07.072

Song W, Zhang K, Chen Z, Hong G, Lin J, Hao C, Zhang S (2018) Effect of xylanase–laccase synergistic pretreatment on physical–mechanical properties of environment-friendly self-bonded bamboo particleboards. J Polym Environ 26(10):4019–4033. https://doi.org/10.1007/s10924-018-1275-7

Su Y, Fang L, Wang P, Lai C, Huang C, Ling Z, Yong Q (2022) Coproduction of xylooligosaccharides and monosaccharides from hardwood by a combination of acetic acid pretreatment, mechanical refining, and enzymatic hydrolysis. Biores Technol 358:127365. https://doi.org/10.1016/j.biortech.2022.127365

Taneda D, Ueno Y, Ikeo M, Okino S (2012) Characteristics of enzyme hydrolysis of cellulose under static condition. Biores Technol 121:154–160. https://doi.org/10.1016/j.biortech.2012.06.104

Tang Y, Geng S, Houtman C, Wu S (2019) Size distribution analysis of microstickies treated by enzyme mixtures in papermaking whitewater. TAPPI J 18(3):183–192. https://doi.org/10.32964/TJ18.3.183

Tolan J, Guenette M (1997) Using enzymes in pulp bleaching: Mill applications. In: Biotechnology in the pulp and paper industry. Springer, pp 289–310

Torres CE, Negro C, Fuente E, Blanco A (2012) Enzymatic approaches in paper industry for pulp refining and biofilm control. Appl Microbiol Biotechnol 96(2):327–344. https://doi.org/10.1007/s00253-012-4345-0

Tyagi C, Singh S, Dutt D (2011) Effect of two fungal strains of *Coprinellus disseminatus* SH-1 NTCC-1163 and SH-2 NTCC-1164 on pulp refining and mechanical strength properties of wheat straw-AQ pulp. Cellul Chem Technol 45:257

Walia A, Mehta P, Guleria S, Shirkot CK (2015) Modification in the properties of paper by using cellulase-free xylanase produced from alkalophilic *Cellulosimicrobium cellulans* CKMX1 in biobleaching of wheat straw pulp. Can J Microbiol 61(9):671–681. https://doi.org/10.1139/cjm-2015-0178

Wang Q, Liu S, Yang G, Chen J, Ni Y (2015) Cationic polyacrylamide enhancing cellulase treatment efficiency of hardwood kraft-based dissolving pulp. Bioresour Technol 183:42–46. https://doi.org/10.1016/j.biortech.2015.02.011

Wang M, Ye Y, Li X, Chen H, Zhao J (2017) Utility of co-expressed alkali-tolerant endoglucanase and xylanase in ameliorating wheat straw chemical pulp properties. Cellulose 24(5):2299–2311

Wang F, Zhang X, Zhang G, Chen J, Sang M, Long Z, Wang B (2018) Studies on the environmentally friendly deinking process employing biological enzymes and composite surfactant. Cellulose 25(5):3079–3089

Wang Q, Fu X, Liu S, Ji X, Wang Y, He H, Yang G, Chen J (2020) Understanding the effect of depth refining on upgrading of dissolving pulp during cellulase treatment. Ind Crops Prod 144:112032. https://doi.org/10.1016/j.indcrop.2019.112032

Wei S, Liu K, Ji X, Wang T, Wang R (2021) Application of enzyme technology in biopulping and biobleaching. Cellulose 28(16):10099–10116

Woolridge EM (2014) Mixed enzyme systems for delignification of lignocellulosic biomass. Catalysts 4(1):1–35. https://doi.org/10.3390/catal4010001

Wu H, Cheng XB, Zhu Y, Zeng W, Chen G, Liang Z (2018) Purification and characterization of a cellulase-free, thermostable endo-xylanase from Streptomyces griseorubens LH-3 and its use in biobleaching on eucalyptus kraft pulp. J Biosci Bioeng 125(1): 46–51

Xiang W, Jianhua M, James T (2006) System for control of stickies in recovered and virgin paper processing (US20060048908A1). Enzymatic Deinking Tech LLC

Xu H, Scott GM, Jiang F, Kelly C (2010) Recombinant manganese peroxidase (rMnP) from *Pichia pastoris*. Part 1: Kraft pulp delignification. Holzforschung 64(1):137–143. https://doi.org/10.1515/HF.2010.017

Yang M, Cai J, Wang C, Du X, Lin J (2017) Characterization of endo-β-mannanase from *Enterobacter ludwigii* MY271 and application in pulp industry. Bioproc Biosyst Eng 40:35–43. https://doi.org/10.1007/s00449-016-1672-z

Yang S, Liu M, Long L, Zhang R, Ding S (2018) Characterization of a cutinase from *Myceliophthora thermophila* and its application in polyester hydrolysis and deinking process. Process Biochem 66:106–112. https://doi.org/10.1016/j.procbio.2017.11.021

Yang S, Yang B, Duan C, Fuller DA, Wang X, Chowdhury SP, Stavik J, Zhang H, Ni Y (2019) Applications of enzymatic technologies to the production of high-quality dissolving pulp: a review. Biores Technol 281:440–448. https://doi.org/10.1016/j.biortech.2019.02.132

Yang M, Li J, Wang S, Zhao F, Zhang C, Zhang C, Han S (2023) Status and trends of enzyme cocktails for efficient and ecological production in the pulp and paper industry. J Clean Prod 418:138196

Yin CF, Xu Y, Deng SK, Yue WL, Zhou NY (2021) A novel esterase DacA(pva) from *Comamonas* sp. strain NyZ500 with deacetylation activity for acetylated polymer polyvinyl alcohol. 87, pp e03016–e03020

Zhang ZJ, Chen YZ, Hu HR, Sang YZ (2013) The beatability-aiding effect of *Aspergillus niger* crude cellulase on bleached simao pine kraft pulp and its mechanism of action. BioResources 8:5861–5870

Zhang Z, Lan D, Zhou P, Li J, Yang B, Wang Y (2017) Control of sticky deposits in wastepaper recycling with thermophilic esterase. Cellulose 24(3):311–321

Zheng H, Liu Y, Liu X, Han Y, Wang J, Lu F (2012) Overexpression of a *Paenibacillus campinasensis* xylanase in *Bacillus megaterium* and its applications to biobleaching of cotton stalk pulp and saccharification of recycled paper sludge. Bioresour Technol 125:182–187. https://doi.org/10.1016/j.biortech.2012.08.101

Zhu B, Wang D, Wei N (2022) Enzyme discovery and engineering for sustainable plastic recycling. Trends Biotechnol 40(1):22–37. https://doi.org/10.1016/j.tibtech.2021.02.008

Chapter 4
Chimeric Enzymes

Abstract This chapter provides an in-depth examination of chimeric enzymes, emphasizing their design, engineering, and potential to transform catalytic processes in the pulp and paper industry. It begins by outlining the fundamental principles behind chimeric enzyme construction, wherein functional domains from different parental enzymes are fused to create multifunctional biocatalysts with enhanced stability, activity, and substrate specificity. Various design strategies—including domain fusion, linker optimization, modular rearrangement, and rational or semi-rational engineering—are discussed to demonstrate how tailored architectures improve catalytic performance under industrial conditions. The chapter also highlights advanced enzyme-engineering approaches such as directed evolution, site-specific mutagenesis, and computational modeling, which further refine chimeric constructs. Practical benefits, including improved thermostability, reduced chemical dependence, and enhanced delignification, are explored alongside emerging applications in biobleaching, deinking, refining, and pitch control. Forward-looking perspectives emphasize the role of chimeric enzymes in enabling greener, more efficient, and sustainable paper production within a circular bioeconomy framework.

Keywords Chimeric enzymes · Domain fusion · Enzyme engineering · Directed evolution · Linker optimization · Biobleaching · Deinking · Pulp and Paper biotechnology · Catalytic enhancement · Sustainable processing

4.1 Chimeric Enzymes: A New Frontier in Protein Engineering

Chimeric enzymes, developed through the process of protein engineering, are becoming increasingly significant in the pulp and paper sector. These enzymes, engineered to possess multiple functions, provide benefits over conventional enzyme mixtures utilized in operations such as pulping, bleaching, and deinking. Chimeric enzymes can enhance operational efficiency, minimize chemical consumption, and improve the overall quality of paper products. A chimeric enzyme is a kind of protein

P. Bajpai, *Chimeric Enzymes Showing Promise in Pulp and Paper Sector*,
SpringerBriefs in Molecular Science, https://doi.org/10.1007/978-3-032-18003-2_4

that is created when two or more different genes combine to produce a translational product. The designation 'chimeric' or 'fusion protein' pertains to the protein generated from the genetic fusion, which entails the incorporation of various segments from multiple genes. Protein engineers often describe these chimeric proteins as "hybrids" because they result from the partial or complete fusion of two or more proteins, which can originate from different sources, serve various functions, or occupy distinct locations within cells. There are two types of chimeric proteins: (1) natural and (2) synthetic or artificial.

4.1.1 Chimeric Enzymes Found in Nature

Because they promote the evolution of protein families, several naturally occurring chimeric proteins are essential to the process of speciation. Typically, these proteins exhibit additional functionalities while preserving the original properties of their constituent proteins. The evolution of protein families is driven by various mechanisms of gene reorganization, like horizontal gene transfer, gene duplication, exon shuffling, chromosome translocation, alternative splicing, noncoding regions' transformation into coding sequences, and the movement of transposable elements. Proteins with altered or completely new functions may arise from this genomic reorganization, and in certain cases, subsequent gene reorganizations may produce new phenotypes. BCR-ABL1, an oncogene linked to chronic myeloid leukemia, is among the most thoroughly researched chimeric proteins found in nature (Rogers and Hartl 2012; Long 2000; Graveley 2001; Shapiro 2002; Piatkowska et al. 2013; Press et al. 2013).

The BCR-ABL1 fusion protein initiates the formation of the Philadelphia (Ph) chromosome. A fusion protein that combines lipoxygenase and heme peroxidase has been found in the Caribbean Sea whip coral *Plexaura homomalla*. These two enzymes work together to make allene oxide, a precursor to substances that resemble prostaglandin (Koljak et al. 1997). The jingwei (jgw) gene family in *Drosophila*, which is crucial for hormone metabolism, is another pertinent illustration of a chimeric gene (Zhang et al. 2004; Long and Langley 1993).

The male-specific *Drosophila* nuclear transport factor-2-related (Dntf-2r) gene was created by retroposing the parental Dntf-2 gene (Betran and Long 2003). There are more examples of spontaneously occurring fusion proteins in the PvuII type Niabella soli strain DSM 19,437 Restriction-modification (R-M) system. The protective methylation of DNA in the newly formed host cell is made possible in this system by the combination of the restriction endonuclease and its transcriptional regulator C protein (Liang and Blumenthal 2013). Furthermore, proline production depends on the bifunctional enzyme pyrroline-5-carboxylate synthase (Turchetto-Zolet et al. 2009).

4.1.2 Artificially Created or Recombinant Chimeras

Protein engineering represents a highly intriguing area of research, characterized by its extensive and profound nature, indicating that our understanding of this field remains limited. Many approaches and strategies have been developed since the 1980s to elucidate the nature, structure, including the determination of active sites, and catalytic functions of specific proteins. Each protein possesses unique functional domains that enable it to perform its designated catalytic activities. While a protein may effectively catalyze biochemical reactions, it may not necessarily exhibit optimal thermostability or high temperature efficiency. Such limitations can arise from factors such as pH levels, metal inducers, or other components of the reaction mixture or conditions, which may hinder a single enzyme's ability to meet the required demands. To enhance enzyme capabilities, the concept of chimeric enzymes has emerged. This idea is crucial for locating important amino acids at the active site or functional areas within polypeptides that are involved in catalytic reactions. The two main approaches used in the creation of chimeric enzymes are directed evolution and rational design (Bornscheuer and Pohl 2001). The result of much research into DNA–protein interactions is the formation of fusion or chimera proteins. Initially, protein engineering focused on the identification and characterization of individual enzymes; however, advancements in technology now allow for the simultaneous analysis of multiple enzymes. As a result, fusion proteins can be produced more automatically, effectively reducing compatibility issues when two proteins are utilized together in a reaction mixture. Presently, designer enzymes are produced based on specific process requirements. Chimeric protein synthesis can be accomplished in a methodical manner by fusing the genes corresponding to target proteins, either with or without linkers, utilizing the technique of recombinant DNA and bioinformatics techniques.

The process of combining the corresponding genes of the target protein, whether with a linker or not, involves cloning them into the correct expression vector and subsequently introducing the vector into an appropriate host. Yeast cells, bacterial cells like *Escherichia coli*, and mammalian cell lines like HEK293 (human embryonic kidney 293) and CHO (Chinese hamster ovary) are frequently utilized for producing chimeric proteins. Additionally, any plant cell may be employed for this purpose, and at times, insect cells are also used. Chimeric proteins' altered catalytic properties are shown in vitro, whereas their phenological or physiological role is expected in vivo. A functional form of the fusion protein is created following the screening of gene combinations. Two or more proteins can fuse to form a bi, tri, or multifunctional protein, or the functional and catalytic domains of many proteins can be inserted into a single peptide (Fig. 4.1) (Bulow and Mosbach 1996; Liu et al. 2018a).

This domain insertion can occur as a single instance (a one-time insertion of a domain) or as multiple instances (involving more than one domain insertion). Additionally, multiple insertions can be categorized into three types: (1) nested insertion (the process of inserting a new domain into an intermediate domain), (2) two-domain insertions, and (3) three-domain insertions (Fig. 4.2).

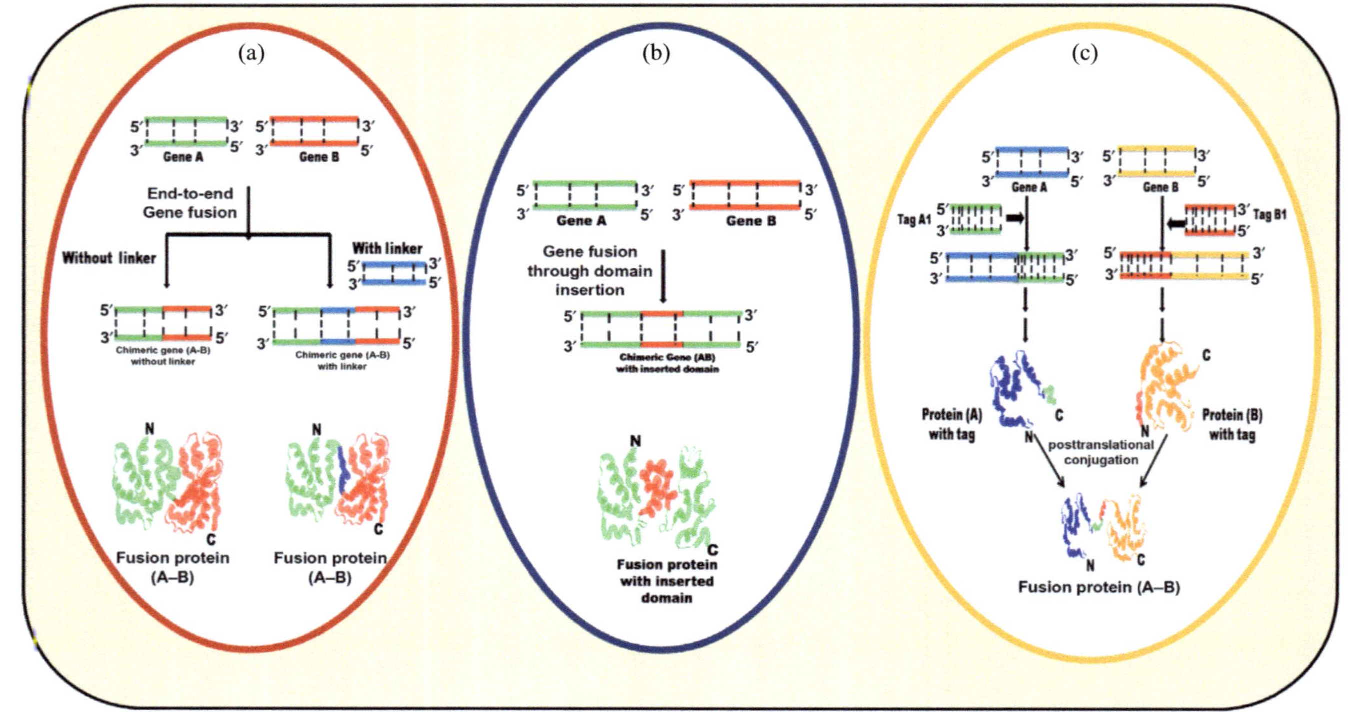

Fig. 4.1 A schematic illustration of the three techniques for fusing domains or proteins: posttranslational conjugation (**c**), domain insertion (**b**), and end-to-end gene fusion (**a**). Jaduan et al. (2020). Reproduced with permission

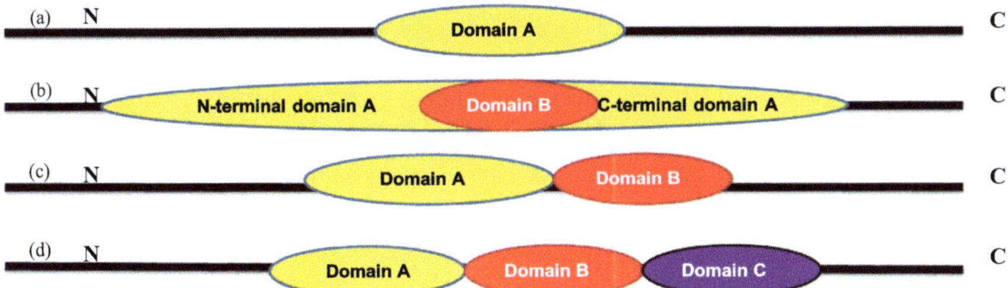

Fig. 4.2 A description of the various forms of domain insertions in proteins: **a** single insertion, **b** nested insertion, **c** two-domain insertion, and **d** three-domain insertion. N and C indicate the N- and C-terminal ends of the parent domain or protein, respectively. Jaduan et al. (2020). Reproduced with permission

A variety of enzymes display multifunctional traits, and by mimicking natural processes, several successful endeavours have been made in this area. Generally, fusion proteins maintain the characteristics of their original components; nonetheless, there are cases where a decrease or complete loss of the parent proteins' functionality has been noted (Roy 1999; Furtado et al. 2013). Because of its potential applications in business, medicine, and agriculture, fusion enzyme research has significantly increased over the past ten years (Fan and Yuan 2010; Beck and Reichert 2011; Saadat 2017). These multifunctional proteins are designed to combine the beneficial features of various proteins into a single peptide.

A novel fusion protein that possesses both cellulase and xylanase capabilities has recently been applied to the decomposition of lignocellulosic biomass, revealing greater efficiency than treatments using single enzymes. When complex polymers like cellulose and xylan are efficiently broken down, simple sugars like glucose and xylose are produced. These can be utilized to create biofuels and other useful resources. Furthermore, this method aids in the management of agricultural and industrial waste with minimal effort. In the realm of waste paper deinking, the created chimeric enzyme lipase-cutinase (Lip-Cut) has demonstrated superior efficiency compared to either of the individual proteins or their combinations. Fusion proteins are essential in the creation of chimeric therapeutic agents, with many chimeric medications having gained FDA approval and being manufactured by various pharmaceutical firms. Multifunctional chimeric antibodies, especially those built on Fc fusion proteins, are proving to be very effective in treating cancer (Khalili et al. 2017; Anwar et al. 2014; Davidi et al. 2016; Narnoliya et al. 2017; Liu et al. 2018a, b; Walsh 2010; Strohl 2017; Czajkowsky et al. 2012).

Considering the importance of fusion proteins in various fields, it is crucial to explore this area further, as advancements in protein engineering technologies can provide significant advantages in tackling issues like fuel scarcity, disease treatments, genetic disorders, and drug delivery challenges.

While maintaining the original functionalities of their constituent proteins, chimeric proteins frequently exhibit enhanced functions (Gilmore et al. 2020). Enzymes with new or different functions can be produced by genetic rearrangements; in some instances, repetitive gene restructuring results in the formation of unique enzymes (García-Paz et al. 2024). Among the various naturally occurring chimeric enzymes identified, lipoxygenase and hemoperoxidase are two notable examples. Allene oxide, produced by this specific chimeric enzyme, serves as a precursor to substances that resemble prostaglandins (Jadaun et al. 2020). Furthermore, proline production depends on the dual-purpose enzyme pyrroline-5-carboxylate synthase (Al-Khayri et al. 2022). The idea of a chimeric enzyme has been suggested to improve its functional abilities.

This idea is also crucial for figuring out which polypeptide functional sections are involved in the catalytic mechanism (García-Paz et al. 2024). Multiple enzyme activity can now be evaluated simultaneously thanks to recent technological advancements. In order to solve compatibility issues that occur when two enzymes are utilized simultaneously in a procedure, a highly automated technique for producing chimeric enzymes has been created. Nowadays, enzymes that have been engineered are purposefully made to satisfy the requirements of this procedure. Chimeric enzymes can be generated using bioinformatics techniques and recombinant DNA technology. In this process, the genes of the desired enzyme are combined, with or without a linker, and inserted into a suitable expression vector before being transferred to a compatible host.

Because of its numerous applications in industry, health, and agriculture, chimeric enzyme development has attracted a lot of attention in the past ten years (Dahiya et al. 2024). A chimeric protein exhibiting both cellulases and xylanases has been utilized to decompose lignocellulosic biomass.

Table 4.1 illustrates synthetic chimeric enzymes that exhibit enhanced physico-chemical properties compared to their native counterparts (García-Paz et al. 2024). It is evident from Table 4.1 that chimerization can influence the stability of enzymes in terms of pH and temperature. Nevertheless, there is a scarcity of research employing this approach to enhance enzyme stability in the presence of organic or eutectic solvents, salts and ionic surfactant liquids. Given that the determinants of pH and thermal stability are analogous to those that contribute to enzyme stabilization in solvents, surfactants and salts, or, we perceive this as a promising avenue for future exploration in protein engineering. Figure 4.3 illustrates instances of multidomain chimeric enzymes that occur in nature, encompassing the seven categories of enzymes (García-Paz et al. 2024).

Chimeric proteins can be readily produced in vitro utilizing recombinant techniques by including the structural genes of the respective proteins into an appropriate expression vector. Fusion proteins can be divided into two basic groups (Fig. 4.4). In the first group, the promoter at the 50-terminus of the resulting structural gene as well as the translational 30-terminus of the original gene are removed. The two genes are subsequently ligated in-frame, either with or without an intervening linker, and produced in the suitable host. The most commonly utilized hosts include bacteria like *Escherichia coli* and two mammalian cell lines: HEK293 (human embryonic kidney-293) and CHO (Chinese hamster ovary). Additionally, other mammalian cell lines, plant cells and insect cells have also been utilized. A protein's fusion might take place at one or both ends. For every fusion construct, it is vital to determine which end is more beneficial for the protein's biological function. Complete structural genes or short synthesized oligonucleotides can be the DNA molecules targeted for fusion.

Table 4.1 Chimeric synthetic enzymes demonstrating superior physicochemical properties relative to native enzymes

Enzyme	EC number	Organism	Domain added	Domain origin	Improved property	References
Laccase	EC 1.10.3.2	*Pleurotus ostreatus*	Complete class I hydro- phobin Vmh2	*Pleurotus ostreatus*	Increased immobilization yield	Sorrentino et al. (2019)
Glicosylhidroolase	EC 3.2.1	*Thermotoga maritima*	Dockerin domain	*Piromyces finnis*	Thermostability	Gilmore et al. (2020)
Glucoamylase GATE	EC 3.2.1.3	*Talaromyces emersonii* Ld418 TE	Starch-Binding Domain of glucoamylase GAA1	*Aspergillus niger* Ld418A1	Enzyme activity pH stability	Chen et al. (2022)
Xylanase	EC 3.2.1.8	*Thermobacillus xylani- lyticus*	N-terminal of GH11 xylanase	*Neocallimastix patri- ciarum*	Wider substrate specificity	Song et al. (2014)
Xylanase A AnxA	EC 3.2.1.8	*Aspergillus niger*	N-terminal of xylanase A TfxA	*Thermomonospora fusca*	Thermostability Catalytic activity	Sun et al. (2005)
Xylanase	EC 3.2.1.8	*Thermobacillus xylani- lyticus*	N-terminal of GH11 xylanase	*Neocallimastix patri- ciarum*	Wider substrate specificity	Song et al. (2014)
Xylanase A AnxA	EC 3.2.1.8	*Aspergillus niger*	N-terminal of xylanase A TfxA	*Thermomonospora fusca*	Thermostability Catalytic activity	Sun et al. (2005)
β-xylanase	EC 3.2.1.8	*B. subtilis*	A complete β-xylosidase	*B. subtilis*	Substrate cleavage rate Optimum temperature Thermostability	Diogo et al. (2015)

(continued)

Table 4.1 (continued)

Enzyme	EC number	Organism	Domain added	Domain origin	Improved property	References
Cyanide dehydratase	EC 4.2.1.66	*B. pumilus*	C-terminal of cyanide	*P. stutzeri*	Thermostability Optimum pH Operational pH range	Crum et al. (2015)
Nitrile hydratase	EC 4.2.1.84	*P. putida* NRRL-18,668	N-terminal of nitrile C-terminal of nitrile hydratase	*Comamonas testeroni* 5-MGAM-4D *P. thermofila* JCM3095	Enhancement in themostability More tolerant to high-concentration product Increase in activity	Cui et al. (2014)

García-Paz et al. (2024). Licensed under a Creative Commons Attribution 4.0 International License

MULTIDOMAIN CHIMERIC ENZYMES
1. Oxidoreductases
Cytochromes
P450_BM-3 system
P450
CPR
2. Transferases
Fructansucrases
InlA
N-terminal ASR identity
Catalityc FTF identity
C-terminal ASR identity
3. Hydrolases
Maltooligosaccharide-forming amylases
MFAses
Catalytic domain
SBD1
SBD2
4. Lyases-pectate
Lyases
PelA
N-terminal PelB identity
Catalityc PelI - PelK identity
Dockerin
5. Isomerase
Disulfure isomerase
PDI
Trx identity
Inactive domain
Inactive domain
trx identity
6. Ligases
Formylglycinamide ribonucleotide amidotransferase
PurL
N-terminal PurS identity
Linker
PurM identity
Glutamine domain
7. Translocase
Recruitment of outd-membrane decarboxylase domain pumps
OadA

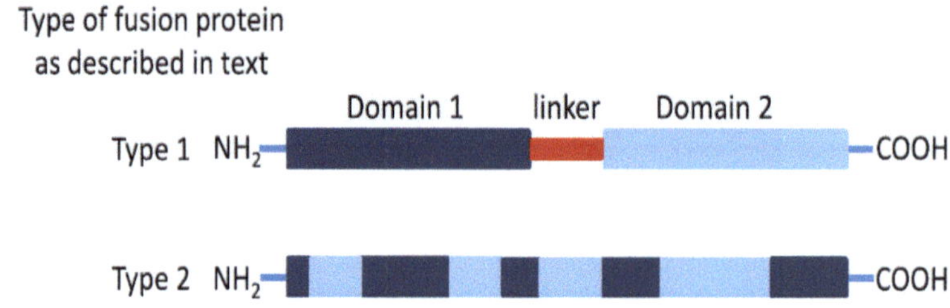

Fig. 4.4 Two fundamental categories of fusion, or chimeric, proteins. The first category comprises two proteins or protein subunits that are fused end-to-end, typically connected by a linker. The second category involves the interspersion of amino acids from both donor proteins within the resulting fusion protein product. Strohl (2017). Reproduced with permission

The second category of fusion proteins, referred to as chimera proteins, is created by carefully intertwining the sequences of two similar proteins to form a novel protein that frequently exhibits unique activity. In contrast to the first category of fusion proteins, this group typically demonstrates a singular type of unique activity rather than two separate activities. With the increasing number of genomes and genes that have been sequenced, along with the application of polymerase chain reaction (PCR) to retrieve genes in their entirety, there exist numerous combinations that could lead to the fusion of structural genes from various sources. Given the nearly limitless combinations of fusion partners, this method has emerged as a valuable and versatile tool across many domains of biochemistry and biotechnology.

4.2 Design of Chimeric Enzymes to Enhance Stability, Activity, and Specificity

4.2.1 *Improvement in Stability*

Enzymes employed in industrial applications must exhibit robust stability under a wide range of harsh conditions, including the presence of organic solvents, high concentrations of salts, and various surfactants. Additionally, they are often required to function effectively at elevated temperatures and across extreme pH ranges, which can significantly challenge their structural integrity and catalytic efficiency (Illanes et al. 2012). One of the most critical biological attributes of multidomain proteins in this context is their inherent thermostability. This property can be enhanced through several molecular strategies, such as reinforcing intermolecular interactions, introducing disulfide bridges to increase structural rigidity, or modifying amino acid residues in regions prone to unfolding or denaturation. These targeted alterations help maintain the protein's functional conformation under stress. Furthermore, numerous studies have shown that enzymes possess the remarkable ability to reorganize or adapt their domain architecture, thereby improving their overall stability and resilience in demanding industrial environments.

As demonstrated by *Bacillus pumilus's* cyanide dehydratase (EC 4.2.1.66), which improved both its thermostability and pH tolerance when its homologous C-terminal domain (56 aa) of the thermostable *P. stutzeri* cyanide dehydratase was substituted for the C-terminal domain, the stability improvement can be seen in multiple characteristics. The chimeric enzyme's half-life was enhanced by 17 times, and pH 8–9 was the ideal range. While the chimeric enzyme exhibited a greater affinity for the substrate, wild-type cyanide dehydratase was inactive at this pH. According to this study, the extended half-life of the chimeric enzyme was caused by oligomerization that was triggered by the C-terminal domain (Crum et al. 2015). Proteins are maintained in a stable state by intermolecular forces like π^* interactions, charge-charge interactions, hydrophobic interactions, disulfide bonds, and hydrogen bonds. Energy-wise, hydrophobic areas are the most stable, even though hydrogen bonds are the primary driver for the development of protein structures. According to some authors, proteins have extremely compact interiors, meaning that their molecules are located at half the distance compared to those in a drop of water. Protein shape can therefore be altered by minute variations in atom displacement, which can either increase or decrease the protein's stability (Nick Pace et al. 2014). Chimerization can alter the temperature and pH stability at which enzymes stay stable. Few research, nevertheless, have employed this tactic to increase enzyme stability in organic or eutectic solvents, salts, and ionic surfactant solutions. We see this as a potential avenue for future protein engineering research because the determinants of pH and temperature stability are comparable to those involved in the stabilization of enzymes in solvents, salts, or surfactants.

The enzymes employed within the pulp and paper sector often exhibit low stability when subjected to elevated temperatures and extreme pH levels. One possible strategy to address these stability issues is the creation of chimeric enzymes. A chimeric enzyme that combines GH10 xylanase and GH62 arabinofuranosidase has recently demonstrated a significant increase in stability when calcium ions are present (dos Santos et al. 2024). A highly desired biological characteristic in multidomain proteins is the thermostability attained by strengthening weak areas through amino acid changes, disulfide bond formation, or increased intermolecular forces. Numerous studies have shown that domains can be effectively moved via enzymes to boost stability (Yin et al. 2013). For example, the native enzymes were 1.5 times less active than a chimeric xylanase-laccase that was utilized for biological bleaching. In addition, after being subjected to 60 °C for two hours, this chimeric enzyme sustained 70% of its activity, while the parent enzymes only managed to maintain 40 and 45% of their activity in the same environment (Saadat 2017). By substituting it with a thermostable alternative, a straightforward modification in the domain can be achieved.

Prior research has demonstrated that the stability of xylanases is influenced by their N-terminal domains (EC 3.2.1.8). To increase the thermostability of mesophilic xylanases, several researchers have partially or completely substituted thermophilic areas for their N-terminal domains. As a result, the enzymes now have up to 10 °C more thermostability than their original forms. Protein stabilization depends on intermolecular forces like π^* interactions, charge-charge interactions, hydrophobic interactions, salt bridges, disulfide and, hydrogen bonds (Yin et al. 2013; Sun et al. 2005).

Accordingly, the development of a thermostable enzyme depends on the creation of salt bridges and disulfide linkages (Yang et al. 2024). A mesophilic enzyme can also be made thermophilic by N-terminal substitution and molecular dynamics simulations.

GH5 endoglucanase and β-1,4-glucosidase fused to form a chimeric enzyme (F194A mutant), which showed greater catalytic efficiency, a lower Km (0.02 mM pNPG) suggesting increased selectivity, and a higher melting temperature (Tm) of 78 °C (Nath et al. 2019). In another research, six chimeric enzymes that target either the C- or Nr terminal were produced by combining naturally occurring β-glucosidase from *Clostridium cellulovorans* with xylanase-cellulase from *Clostridium thermocellum*. For the formation of amorphous cellulose, this chimeric enzyme demonstrated enhanced pH and temperature stability in addition to a 2.3-fold improvement in catalytic efficiency towards carboxymethyl cellulose (Chen et al. 2019).

4.2.2   Improvement in Activity

For effective bioconversion, it is crucial that the optimal biocatalyst has a high specificity constant (kcat/Km) or a high turnover number (kcat). To improve enzymatic performance, several strategies have been used, such as directed evolution, the use of organic solvents, and immobilization. Chimeric enzymes may be used in the development of effective biocatalysts. It is possible to modify these chimeric enzymes, which integrate multiple catalytic processes into one, to create new products. These pairings usually result in improved biocatalyst activity and higher catalytic efficiency in bioprocesses (García-Paz et al. 2024).

Adding particular modules to single-domain enzymes is another method of increasing the activity of chimeric enzymes. Although they are not catalytic domains per se, carbohydrate-binding modules have a unique fold that increases enzymatic activity and exhibit a high affinity for polysaccharides (Carli et al. 2022). By facilitating interactions between the enzyme and the substrate, these modules raise the concentration of the substrate locally and enhance enzyme activity. A 30% increase in xyloglucan hydrolysis was achieved, for example, when *C. thermocellum* xyloglucan-specific CBM44 (EC 3.2.1.151) and *Aspergillus niveus* GH12 XEGA were fused to create a chimeric enzyme (CBM44-XEGA) with a kcat 1.25 times higher in comparison to the original enzyme. Interestingly, the carbohydrate-binding module did not alter the enzyme's catalytic mechanism because the hydrolysis product profile stayed the same (Furtado et al. 2015). Additionally, by connecting the carboxyl terminus of XynB, a thermostable single-domain xylanase from the *Thermotoga maritima* family 10, with the carbohydrate-binding module from *Streptomyces thermoviolaceus* STX-II family 2b, a 1.7-fold improvement in kcat for soluble xylan extracted from birchwood was achieved. The chimeric enzyme XYN-TmCBM9-1_2, which fuses the carbohydrate-binding module TmCBM9-1 2 from *T. maritima* with the xylanase XYN from *A. niger*, serves as another significant example. The incorporation of insoluble xylans from birchwood enhanced this hybrid enzyme's xylanase activity by 4.2 times. (García-Paz et al. 2024).

Rosetta's active site design is a popular method for enhancing enzyme performance. The first step in this process is to visualize a perfect hypothetical active site (enzyme) that has every functional group required to support the intended reaction. A database of protein structures was searched in order to locate a protein scaffold that could bolster this theory.

A number of software programs have been created especially for the de novo synthesis of enzymes, like SABER, Rosetta and ORBIT (Planas-Iglesias et al. 2021a, b; Leman et al. 2020; Ferreira et al. 2022). The process for identifying hotspots differs depending on the improved certain enzymatic properties.

In comparison to the original enzyme, a chimeric enzyme consisting of GH11 xylanase and a xylose-binding protein with an altered active site exhibited three times the activity and enhanced thermal stability. Furthermore, a thermostable scaffold and Rosetta-engineered β-1,4-xylanase showed a 1.3-fold increase in activity in comparison to the wild-type enzyme (Moretti et al. 2016).

The Rosetta enzyme design methodology, which mainly concentrates on enhancing the rigidity of flexible regions, has also been utilized to augment the stability of inherently unstable enzymes. Specific areas within these enzymes were modified through mutagenesis and subsequently subjected to re-evaluation. This method looks at characteristics of the enzyme–substrate complex, like specificity, selectivity and catalytic activity, by highlighting amino acids that have direct interactions with the ligand. Among the notable software tools developed for this purpose are FoldX, Rosetta, ORBIT, and PELE (Zhao et al. 2021). Sequence conservation analyses are typically carried out to discover residues that might be essential to an enzyme's overall stability and activity. Tools like the HotSpot Wizard were used in these studies (Sumbalova et al. 2018). The following residues are targeted for mutation in a variety of enzymes, including endoglucanases, which undergo de novo enzyme design, specifically through the bridging of internal disulfide bonds: DSB3: Tyr171Cys/Leu201Cys and DSB2: Ser127Cys/Ala165Cys. The resultant mutant enzyme exhibits improved lignocellulose breakdown and increased thermostability (Bashirova et al. 2019). Additionally, using the SpyTag/SpyCatcher system, isopeptide bond-mediated molecular cyclization was used to create a chimeric enzyme that combines mannanase and xylanase (Luo et al. 2024). This chimeric enzyme demonstrated resilience under freeze–thaw cycles, thermostability, and resistance to ionic environments (Gao et al. 2019).

4.2.3 Merging and Altering the Specificity

The obstacles associated with the development of enhanced biocatalysts must be tackled by researchers to facilitate the synthesis of novel compounds and enhance bioprocesses. Another approach to accomplish this involves the design of biocatalysts through the combination or chimerization of enzyme specificities. The integration of various enzyme domains can alter selectivity and introduce new biosynthetic functions (García-Paz et al. 2024). Furthermore, additional characteristics of the enzyme, such as its activity or selectivity, can be enhanced through combined functionality.

Using a 26-residue linker, a β-xylanase (26 kDa) from *Bacillus subtilis* was joined with the catalytic domain of a β-xylosidase (60 kDa) from the same organism to form an enhanced thermostable bifunctional chimera. After a duration of 45 h of incubation at 45 °C, the preferred temperature for β-xylosidase activity was elevated by 5.3 °C, whereas the wild-type enzyme only maintained 30% of its activity. Furthermore, three different types of xylose were generated by the chimeric enzyme (Hegazy et al. 2019). Another example is the creation of four chimeric enzymes using β-1,4-endoxylanase (Xyl11D) and β-1,4-endoglucanase (Endo5A) from *Paenibacillus* sp. ICGEB2008 (Adlakha et al. 2011). The enzymes were altered, and a glycine-serine linker was used to create chimeric enzymes. It was discovered that the chimeric enzymes' activity were 1.3–2.3 and 0.5–1.6 times higher than those of Xyl11D and Endo5A, respectively. One of the chimeras had the greatest quantities of endoglucanase (1070 U μmol min) and xylanase (899 U μmol min).

Chimeric enzymes are created by combining two enzymes that promote consecutive operations. The next enzyme uses the substrate to create the final product after the first enzyme transforms a substrate into a product. Additionally, when compared to the solo enzyme, the β-glucosidase and endoglucanase fusion showed better selectivity (0.267 mg/ml) for its substrate pNPG. The development of chimeric enzymes serves as a method to assess the additional processes necessary for the synthesis of metabolites and other compounds relevant to medicinal chemistry (Jadaun et al. 2020).

By employing the previously mentioned techniques, chimeric enzymes were developed using a range of computational tools and online servers. The nucleotide sequences were examined and compiled with the SnapGene software (http://sna pgene.com). The theoretical molecular weight and extinction coefficient of a recombinant protein are calculated using ProtParam (https://web.expasy.org/protparam/). Sequences of proteins were aligned using BLASTp (http://www.ncbi.nlm.nih.gov/BLAST/). ClustalW (https://www.genome.jp/tools-bin/clustalw) was utilized to align the amino acid sequences of the enzymes, while ESPript 3.0 (http://espript.ibc p.fr/) was employed to visualize the outcomes. The SignalP program (http://www.cbs.dtu.dk/services/SignalP/) was employed in the prediction of signal peptides. The substrate-binding module's possible segregation sites and catalytic domain of the target enzyme were predicted using the GinZu technique.

Design considerations for chimeric enzymes are presented in Table 4.2. Table 4.3 presents chimeric enzyme constructs and functional enhancements.

Table 4.2 Design considerations for chimeric enzymes

Design element	Options	Impact on functionality
Domain selection	Catalytic, CBM, redox, regulatory	Determines synergy and expression compatibility
Linker type	Rigid, flexible, protease-sensitive	Affects folding, domain orientation, and activity
Expression host	*E. coli*, yeast, fungi, mammalian cells	Influences glycosylation, folding, and stability

Table 4.3 Chimeric enzyme constructs and functional enhancements

CBM-Enhanced Chimeric Constructs

Fusion of CBMs with catalytic domains has yielded significant improvements in activity:

- **CBM44-XEGA**: Fusion of *Aspergillus niveus* GH12 XEGA with Clostridium *thermocellum* CBM44 (EC 3.2.1.151) increased kcat by 1.25-fold and xyloglucan hydrolysis by 30%, without altering the hydrolysis profile
  - **XynB-STX-II**: A fusion of thermostable xylanase XynB from *Thermotoga* maritima with CBM from *Streptomyces thermoviolaceus* (family 2b) resulted in a 1.7-fold increase in kcat against soluble birchwood xylan
- **XYN-TmCBM9-1_2**: A hybrid of xylanase XYN from *A. niger* and CBM9-1_2 from T. maritima showed a 4.2-fold increase in activity on insoluble birchwood xylan

Bifunctional and Thermostable Chimeras

Bifunctional chimeras demonstrate enhanced thermostability and catalytic synergy:

- A chimera of β-xylanase (26 kDa) and β-xylosidase (60 kDa) from *Bacillus subtilis*, linked via a 26-residue linker, retained 80% activity after 45 h at 45 °C, compared to 30% for the wild-type, and tripled xylose production
- Four chimeras combining Endo5A and Xyl11D from *Paenibacillus* sp. ICGEB2008, linked via glycine-serine sequences, exhibited 0.5–1.6 × and 1.3–2.3 × activity increases over Xyl11D and Endo5A, respectively. One construct achieved 899 U µmol^{-1} min^{-1} (xylanase) and 1070 U µmol^{-1} min^{-1} (endoglucanase)
- Fusion of β-glucosidase with endoglucanase improved specificity for pNPG (0.267 mg/ml), supporting its utility in metabolite synthesis

Multifunctional Constructs with Enhanced Catalysis

Multifunctional chimeras exhibit improved catalytic efficiency and temperature optima:

- Fusion of CcBglA β-glucosidase (P172L) with bifunctional cellulase/xylanase from *C. thermocellum* increased the temperature optimum by 10 °C
- A chimera of F194A mutant β-1,4-endoglucanase and β-glucosidase showed a fivefold increase in catalytic efficiency
- β-xylosidase–endoxylanase chimeras (S160N-R333Q/H) enhanced corn cob hydrolysis by 32.8%

Solvent and Metal Ion Tolerance

Chimeric enzymes demonstrate superior stability in harsh conditions:

- β-glucosidase–xylanase increased reducing sugar release from 12.97 g/l to 19.69 g/l
- Mannanase–xylanase chimera showed 36.67% higher kcat and improved hydrolysis in ionic solvents
- Xylanase–CBD3a fusion yielded 1.6–3 × more reducing sugars in ethanol/methanol mixtures
- Laccase–β-1,3–1,4-glucanase chimera increased sugar release by 20% in 80% ethanol
- Endoxylanase–β-xylosidase chimera exhibited 31-fold and 18-fold higher activity under ethanol treatment
- Xylanase–mannanase chimera retained 30% more activity in the presence of metal ions and EDTA, with 50% and 40% activity increases in Cr^{2+}/Cu^{2+} and EDTA, respectively

Substrate Selectivity and Affinity

Enhanced substrate specificity and affinity are key outcomes of chimeric design:

- Feruloyl esterase–endoxylanase chimera improved substrate specificity and catalytic activity by 50%
- Lichenase–xylanase chimera showed broader substrate utilization and a 1.7-fold increase in catalytic efficiency

Fusion of β-1,3–1,4-glucanases from *C. thermocellum* and *B. amyloliquefaciens* improved substrate affinity (Km 1.2 mg/ml vs. 1.5 mg/ml)

Based on Furtado et al. 2015; García-Paz et al. 2024; Hegazy et al. 2019; Adlakha et al. 2011; Jadaun et al. 2020; Chen et al. 2019; Wu et al. 2019; Ariaeenejad et al. 2023; Gao et al. 2019; Wang et al. 2020; Lin et al. 2018; Monica et al. 2024

4.3 Advantages of Utilizing Chimeric Enzymes for Catalytic Processes

4.3.1 Introduction

The selection of highly efficient catalytic proteins is the result of three billion years of evolution. However, these proteins are often not suitable for use in industrial settings (Martins et al. 2020). Although it is feasible to look for extremophile enzymes that are active enough for particular processes, nature has adapted these enzymes to work in complex molecular mixtures, either in cells or in the environments where these organisms live, not in industrial bioreactors. Additional optimization is necessary to attain all the needed qualities (Minshull and Stemmer 1999). Another technique has been utilized in addition to creating recombinant enzymes: translational genetic fusion of proteins with different properties. By creating multifunctional enzymes from a single polypeptide chain, this technique improves regulatory systems and maximizes enzymatic activity (Elleuche 2015).

4.3.2 Definition and Concept of Chimeric Enzymes

Chimeric proteins, also known as fusion proteins, are synthetic molecules formed by merging genetic sequences from two or more distinct proteins or genes. These hybrids are engineered to incorporate functional domains or regions from different parent proteins, resulting in the acquisition of new or modified functions not typically found in nature. Chimeric proteins have been engineered to enhance performance while reducing production expenses.

4.3.3 Applications in Medicine and Biotechnology

The primary purpose of creating chimeric proteins is to leverage the unique properties of individual protein domains for various applications. In scientific research, chimeric proteins serve as essential tools for investigating protein–protein interactions, cellular processes, and signaling pathways. They enable researchers to gain insights into how these functional domains collaborate or operate independently.

In the realm of medicine, chimeric proteins have significant therapeutic applications. For instance, some cancer treatments utilize chimeric monoclonal antibodies, which combine elements from immune system proteins and tumor-targeting antibodies. These chimeric antibodies demonstrate improved efficacy in targeting cancer cells while minimizing side effects.

Biotechnology and pharmaceutical development often rely on chimeric proteins to engineer molecules with specific properties. This approach can lead to the creation of chimeric cytokines with enhanced therapeutic effects or genetically modified enzymes for various industrial processes, such as biofuel production.

Chimeric proteins are also instrumental in vaccine development. Some vaccines, including certain COVID-19 vaccines, incorporate chimeric antigens. These antigens merge parts of the target pathogen with other elements to enhance the immune system's recognition and response, ultimately providing protection against infectious diseases.

In addition to these applications, chimeric proteins can naturally arise due to genetic mutations or rearrangements. In diseases like certain types of cancer, gene fusions can lead to the production of abnormal chimeric proteins that contribute to the disease's progression.

4.3.4 Industrial and Biotechnological Relevance

Numerous studies have shown that chimeric enzymes work better than enzyme cocktails, despite the fact that enzyme combinations are frequently utilized in business (Ariaeenejad et al. 2023). The intermediate can be instantly transferred to the subsequent active site in the appropriate enzyme by a chimera that catalyzes successive processes. Reaction efficiency is so increased since the intermediate is used up before it is distributed throughout the solution (Saadat 2017; Fan et al. 2009a; Guo et al. 2013). The physical and chemical characteristics of two or more fused proteins, such as temperature and pH profile, may be integrated in addition to their synergistic effect and simpler synthesis. Therefore, under the same operating circumstances, fused proteins have biocatalysis that is similarly stable and active (Elleuche 2015; Rizk et al. 2015). Cellulosomes are among the many multi-domain proteins found in nature (Yu et al. 2015; Fan et al. 2009b).

The complexity of the mechanisms of genomic rearrangement has become more apparent as a result of the significant rise in the availability of sequential data. Consequently, a significant amount of an organism's protein complement consists of multi-modular proteins, often known as chimeras (Henikoff et al. 1997). Drawing inspiration from natural processes, bioengineers have been creating synthetic domain fusion techniques to create chimeric proteins (Saadat 2017; Yu et al. 2015; Ribeiro et al. 2011). When compared to the activities of the parent enzymes, the closeness of the fused enzymes' catalytic domains enhances enzymatic activity. Chimeric proteins are designed to have improved stability, specificity, affinity, and expression efficiency. Optimizing enzymes engaged in depolymerization has been demonstrated to be feasible through the development of bifunctional enzymes that work on lignocellulosic materials.

Even if there are several approaches, there are some common guidelines that must be adhered to in order to provide the proteins proper functionality. These rules, which include translation efficiency, fusion flexibility, and the arrangement of the fused structures according to molecular size, structure, and thermostability, are applicable independent of the protein fusion procedure (Saadat 2017; Elleuche 2015). The successful production of chimeras also depends on achieving appropriate catalytic activity in relation to the physical–chemical parameters (such as the optimal pH and temperature) of the different proteins (Ribeiro et al. 2011). In order to optimize the hydrolysis of lignocellulosic biomass, numerous chimeras have been developed. A laccase/glucanase chimera produced via domain insertion was 20% more active on sugarcane bagasse than the combination of these enzymes (Furtado et al. 2013). An endoxylanase/endomannase chimera produced by tandem fusion demonstrated synergistic effect in hemicellulose hydrolysis (Guo et al. 2013). Tandem fusion produced a xylosidase/arabinofuranosidase chimera that released 30% more xylose (Fan et al. 2009a, b, c). A laccase/endoxylanase built via domain insertion significantly improved the endoxylanase domain's resistance to proteolytic attack and thermotolerance (Ribeiro et al. 2011). The chimera created by fusing two complimentary enzymes, xylanase and xylosidase, exhibited a higher optimum temperature and thermotolerance for the xylosidase domain than the parent enzymes, which enhanced the production of xylose (Diogo et al. 2015). The use of synthetically joined proteins for process improvement is common. Domain insertion, post-translational conjugation, and tandem fusion (end-to-end) are the three methods utilized to encourage domain fusion. The development of these tactics is followed by computational prediction of the structures. The procedures discussed above are relevant to multicomponent systems even though they concentrate on the fusion of two proteins.

Numerous chimeric enzymes show improved substrate specificity, product selectivity, catalytic efficiency, and thermostability (Chen et al. 2022; Peng et al. 2018; Parashar and Satyanarayana 2016, 2017; Arsalan et al. 2024; Chen et al. 2019; Du et al. 2021; Gao et al. 2019; Guo et al. 2013; Lin et al. 2018; Monica et al. 2024; Pinheiro et al. 2021. Rizk et al. 2015; Victorino da Silva et al. 2022; Wu et al. 2019). Parashar and Satyanarayana (2016) constructed a chimeric α-amylase by fusing the domains of amylases from *Bacillus acidicola* and *Geobacillus thermoleovorans*, which showed enhanced thermostability and catalytic efficiency.

When enzymes unite to form chimeras, the physical proximity of the resulting catalytic domains may accelerate processes, and chimeric enzymes frequently exhibit superior performance (Gilbert et al. 1998). By combining the functionalities of pre-existing enzymes through structure-based design, hybrid enzymes are produced (Aÿ et al. 1998). Hence, these enzymes (or their fragments) might be useful as building blocks to produce polypeptides with combinations of catalytic activity that aren't seen in nature (Nixon et al. 1998).

Chimeric proteins have been designed to increase substrate affinity for increased biosensor sensitivity (Guan et al. 2007; Dan et al. 2008), create new substrate specificities (Thulasiram et al. 2007; Zhang et al. 2007), and improve stability (Hua et al. 2004; Mandai et al. 2009). It has been demonstrated that developing bifunctional enzymes that interact with lignocellulosic materials is a viable way to enhance the enzymes engaged in a number of activities, such as the breakdown of biomass (Fan et al. 2009c).

4.3.5 Design Considerations and Challenges

The maintenance or enhancement of the parental proteins' catalytic properties is a crucial component in the creation of enzyme chimeras. A popular strategy is the "end-to-end" fusing of the parental enzymes' N and C termini; yet, this method may produce nonfunctional chimeras because of misfolding or limitations brought on by steric hindrance between the domains (Hong et al. 2007). To improve inter-domain flexibility and preserve the parental enzymes' functioning, a loop may occasionally be inserted at the fusion site to get around these limitations (Crasto and Feng 2000; Fan et al. 2009a, b, c).

For the successful creation of enzyme chimeras, the two catalytic activities must be compatible with respect to their physicochemical requirements, such as pH and temperature optima. Two novel multifunctional enzymes that combine xylanase and laccase activity have been designed, synthesized, and characterized by Ribeiro et al. (2011). The introduction of a thermophilic form of the xylanase domain improved the catalytic overlap of the two activities in the chimeras and increased the compatibility between the catalytic functions. In catalytic processes, chimeric enzymes—which are produced by combining various enzymatic activity in a single protein—offer a number of benefits. They support multi-step reactions in a single reaction vessel and allow for the synthesis of novel compounds, which improves biocatalytic efficiency and lowers production costs.

Here is a more comprehensive examination of the benefits of chimeric enzymes in industrial and biotechnological applications:

Enhanced Catalytic Activity: Chimeric enzymes often demonstrate superior catalytic performance compared to their individual, non-fused counterparts. This enhancement is frequently attributed to the synergistic interactions between multiple catalytic domains, which can lead to increased turnover rates, improved reaction kinetics, and greater overall efficiency in substrate conversion.

Increased Stability: The strategic fusion of distinct protein domains can result in enhanced structural integrity and thermal stability of the enzyme. This increased resilience allows the chimeric enzyme to maintain functionality under a broader range of environmental conditions, including fluctuations in temperature, pH, and ionic strength, thereby extending its operational lifespan in industrial processes.

Improved Substrate Specificity: Through rational design and domain selection, chimeric enzymes can be engineered to exhibit highly refined substrate specificity. This precision enables them to selectively recognize and act upon desired substrates while minimizing off-target activity, which is particularly advantageous in complex reaction mixtures or when high product purity is required.

Multi-step Reaction Capabilities: One of the most compelling advantages of chimeric enzymes is their ability to facilitate sequential catalytic steps within a single molecular framework. This multifunctionality allows for the execution of multi-step biochemical transformations in a single reaction vessel, thereby eliminating intermediate purification steps, reducing processing time, and improving overall workflow efficiency.

Novel Product Formation: The integration of diverse catalytic domains within a single enzyme can unlock new reaction pathways, leading to the synthesis of unique chemical entities that are unattainable using conventional mono-functional enzymes. This capability is particularly valuable in the development of novel biomaterials, pharmaceuticals, and specialty chemicals.

Reduced Production Costs: By consolidating multiple enzymatic functions into a single biocatalyst, chimeric enzymes can streamline manufacturing processes and reduce the need for multiple enzyme preparations. This simplification translates into lower raw material costs, reduced labor and equipment usage, and diminished energy consumption, ultimately contributing to more cost-effective production strategies.

4.4   Designing Strategies for Chimeric Enzymes

Chimeric enzymes can combine the catalytic prowess of one enzyme with the substrate specificity, stability, or regulatory features of another. This modular approach mimics evolutionary processes such as gene duplication, recombination, and domain shuffling, enabling the creation of multifunctional proteins that do not exist in nature. Designing chimeric enzymes is an essential strategy for developing biocatalysts with improved or entirely new functional properties. By combining catalytic domains or structural motifs from different parent enzymes into a single hybrid molecule, researchers can tailor specific traits such as substrate specificity, catalytic efficiency, stability under extreme conditions, or synergistic activity toward complex substrates. This rational or semi-rational design approach enables the creation of multifunctional enzymes that can overcome the limitations of natural enzymes and meet the diverse requirements of industrial, environmental, and biotechnological applications. The design of chimeric enzymes requires a multidisciplinary toolkit encompassing molecular biology, structural bioinformatics, computational modeling, and directed evolution.

Rational design is a cornerstone of protein engineering, enabling the deliberate construction of chimeric enzymes based on structural, functional, and evolutionary insights. Unlike random mutagenesis or directed evolution, rational design relies on detailed knowledge of enzyme architecture, domain interactions, and catalytic mechanisms. This chapter explores the key strategies for designing chimeric enzymes

4.4.1 Domain Swapping and Fusion in Chimeric Enzyme Design

The creation of chimeric enzymes through domain swapping and fusion has emerged as a powerful and versatile strategy in modern protein engineering. These approaches exploit the inherent modularity of proteins, enabling researchers to recombine structural and functional elements from diverse biological sources to generate hybrid biocatalysts with tailored characteristics. Such engineered enzymes hold immense promise in industrial biotechnology, pharmaceuticals, and environmental applications, as they can be customized for enhanced stability, modified substrate specificity, or even the acquisition of entirely novel catalytic activities (Bornscheuer et al. 2012; Bornscheuer and Pohl 2001).

Proteins, including enzymes, are typically composed of one or more domains—compact, independently folding units that often correlate with specific functions such as substrate binding, catalysis, or structural stabilization. This modular architecture allows for the exchange or recombination of domains between different proteins without necessarily disrupting the overall folding or catalytic potential of the resulting enzyme (Doolittle 1995). The principles of domain swapping and domain fusion are rooted in the concept that evolutionary processes themselves have frequently generated new enzyme functionalities through recombination and rearrangement of existing protein modules (Bashton and Chothia 2007).

In a domain-swapping approach, a defined region or domain of one enzyme (the recipient) is replaced with a corresponding or analogous domain from another enzyme (the donor), often within the same enzyme family or structural class. The goal is typically to retain the catalytic activity of the parent enzyme while modifying specific properties such as substrate affinity, optimal temperature, solvent tolerance, or reaction kinetics. Domain fusion, by contrast, involves joining two or more entire domains or even whole enzymes into a single polypeptide chain. This can result in multifunctional chimeras that carry out multiple catalytic reactions sequentially or synergistically, often improving reaction efficiency and product yield.

Domain swapping serves as an elegant means of fine-tuning enzyme properties. By exchanging a catalytic or binding domain with one derived from a related enzyme, researchers can manipulate the physicochemical environment of the active site or alter substrate recognition while maintaining the overall protein framework. For example, substituting the catalytic domain of a mesophilic enzyme with that from a thermophilic homolog has yielded chimeric enzymes capable of operating efficiently at elevated temperatures, a feature desirable for industrial processes that require thermal stability (Leemhuis et al. 2009).

Another successful application of domain swapping lies in modifying substrate specificity. Enzymes that share similar catalytic mechanisms but differ in substrate scope can serve as donors and recipients in swap experiments. This approach allows researchers to endow an enzyme with the ability to act on alternative or non-natural substrates without compromising its catalytic efficiency. For instance, domain exchange between different glycosyl hydrolases has been used to alter carbohydrate-binding preferences, generating enzymes with novel activities suitable for biomass conversion.

The success of domain swapping depends critically on maintaining structural compatibility between the exchanged regions. The swapped domains must interact harmoniously at their junction interfaces to preserve proper folding and functionality. Incompatible interfaces may result in misfolded or inactive proteins. Consequently, computational methods such as sequence alignment, homology modeling, and molecular dynamics simulations are employed to assess potential structural conflicts and optimize linker regions (Chung et al. 2009). Experimental validation involves expressing the chimeric enzyme in a suitable host, followed by purification and biochemical characterization to evaluate kinetic constants (Km and Vmax), thermostability, and activity under various conditions.

While domain swapping focuses on replacing specific regions, domain fusion is aimed at creating entirely new multifunctional proteins. In this technique, two or more enzymes—either catalytically related or distinct—are covalently linked, often through flexible peptide linkers. The resulting fusion enzyme can catalyze consecutive reactions, thereby reducing the need for multiple enzymes in a single bioprocess and minimizing the diffusion of intermediates. This substrate channeling phenomenon enhances overall catalytic efficiency and can improve the yield of desired products (Kumar 2020).

A classic example is the fusion of alcohol dehydrogenase and aldehyde dehydrogenase, which allows efficient conversion of alcohols to carboxylic acids within a single protein construct. Similarly, fusion of cellulase and xylanase has been shown to improve the degradation of lignocellulosic biomass by combining complementary hydrolytic activities (Wang et al. 2023). In many cases, fusion also improves the solubility or stability of individual domains, possibly due to structural stabilization through interdomain interactions.

Designing effective fusion enzymes requires careful consideration of linker sequences. Linkers should provide sufficient flexibility to allow each domain to fold independently and function optimally while maintaining spatial proximity to facilitate substrate transfer. Glycine-rich linkers are often used to provide flexibility, whereas proline-rich or helical linkers can maintain rigidity where necessary (Amet et al. 2009). The choice of linker length and composition is usually determined empirically or optimized through computational modeling.

The field of chimeric enzyme design has greatly benefited from advances in computational biology and structural bioinformatics. Tools such as homology modeling, molecular docking, and molecular dynamics simulations help predict the structural compatibility of domains before experimental construction. Sequence alignment and structural superposition analyses can identify conserved motifs and potential junction points for recombination (Arnold et al. 1998). Furthermore, machine learning algorithms are increasingly being used to predict the effects of domain recombination on enzyme stability and activity based on large datasets of known protein structures (Yang et al. 2022).

Experimentally, synthetic biology techniques such as modular cloning, Gibson assembly, and site-directed mutagenesis enable precise construction of chimeric genes encoding domain-swapped or fused proteins. Expression systems—commonly *Escherichia coli, Pichia pastoris,* or *Bacillus subtilis*—are selected based on the solubility and post-translational modification requirements of the engineered enzyme. Following expression, purification and kinetic assays are performed to characterize enzymatic performance. Modern biophysical tools like circular dichroism spectroscopy, differential scanning calorimetry, and X-ray crystallography further aid in understanding how domain recombination affects structural stability and dynamics.

The industrial relevance of domain swapping and fusion lies in their ability to generate enzymes tailored for specific process conditions. Many industrial biocatalytic reactions require enzymes that function in extreme environments—high temperatures, non-aqueous solvents, or alkaline pH. By swapping domains from extremophilic organisms or fusing complementary enzymes, researchers can design catalysts that meet these demanding conditions (Turner 2009).

In biorefinery applications, domain-fused enzymes such as cellulase-xylanase or cellulase-β-glucosidase hybrids have been instrumental in enhancing biomass saccharification efficiency, reducing enzyme loading and cost (Wilson and Kostylev 2012). In pharmaceutical biotechnology, fusion of enzymes involved in multi-step synthesis pathways can lead to one-pot biosynthetic reactions, improving yield and reducing downstream processing. Furthermore, in environmental biotechnology, chimeric oxidoreductases and hydrolases have been developed for efficient degradation of pollutants and industrial wastes (Singh et al. 2016).

Despite their potential, designing functional chimeric enzymes through domain swapping or fusion is not without challenges. Structural incompatibility, improper folding, or interference between domains can lead to inactive proteins. Additionally, predicting the behavior of interdomain interfaces remains a complex task due to the intricate nature of protein folding and dynamics. However, the increasing integration of AI-based predictive models, directed evolution, and high-throughput screening is expected to overcome these barriers.

Looking forward, the synergy between computational design, synthetic biology, and experimental validation will further advance the field. With the growing availability of protein structure databases such as AlphaFold, researchers can now model enzyme architectures with unprecedented accuracy, facilitating rational domain recombination. The ultimate goal is to develop predictive frameworks capable of designing enzymes de novo, optimized for specific industrial functions with minimal empirical testing.

4.4.2 Active Site Preservation in Chimeric Enzyme Design

Active site preservation is a fundamental consideration in the design and engineering of chimeric enzymes. The catalytic efficiency, substrate specificity, and overall functionality of an enzyme are intrinsically dependent on the precise architecture and chemical environment of its active site. When constructing chimeric enzymes through strategies such as domain swapping or domain fusion, careful preservation of the structural and functional integrity of this catalytic region is essential. Properly maintained, the active site enables the engineered enzyme to retain its original catalytic capabilities or acquire new and enhanced functions without loss of activity or selectivity.

The active site of an enzyme represents a uniquely organized microenvironment where catalysis occurs. It typically comprises a set of amino acid residues spatially arranged to facilitate substrate recognition, binding, and conversion into products. These residues work cooperatively to stabilize the transition state, reduce activation energy, and enable efficient turnover of substrates (Baldwin 1999). In addition to the catalytic residues directly involved in chemical transformation, surrounding residues play crucial roles in maintaining proper geometry, polarity, and dynamics of the catalytic core.

The architecture of the active site is often supported by specific secondary and tertiary structural motifs—α-helices, β-sheets, or loop regions—that contribute to maintaining the correct spatial configuration. Even minor perturbations in the positioning of catalytic residues can dramatically affect enzyme activity. For example, displacement of a single side chain involved in proton transfer or substrate orientation can lead to significant reductions in catalytic efficiency or complete inactivation. Therefore, any structural modification during chimeric enzyme construction must avoid distortion of the catalytic site geometry.

Designing chimeric enzymes inherently involves recombining domains from different proteins, often derived from diverse sources or species. While such recombination can endow enzymes with desirable traits—such as enhanced stability, expanded substrate range, or improved tolerance to extreme conditions—it also introduces a major challenge: ensuring structural compatibility between the donor and recipient domains. Misalignment at the domain interface can alter backbone conformations or distort the folding pattern, indirectly perturbing the active site geometry (Bornscheuer et al. 2012).

Conformational coupling between the active site and distal regions of the protein adds further complexity. The catalytic site does not operate in isolation; rather, it is dynamically connected to the overall protein scaffold. Changes introduced through domain fusion or swapping—particularly at interdomain linkers—can propagate conformational strain that affects active site flexibility and accessibility. In some cases, steric hindrance near the catalytic cleft may block substrate entry or interfere with product release, thereby reducing turnover rates. Consequently, maintaining the structural communication between the active site and surrounding domains is as critical as preserving the catalytic residues themselves.

Maintaining the active site's structure and function during chimeric enzyme engineering requires a combination of computational modeling, rational design, and experimental validation.

(a) Structural Alignment and Homology Modeling: Prior to domain recombination, detailed structural analysis of both donor and recipient enzymes is conducted to identify regions of homology and potential junction points. Sequence alignment and homology modeling help locate conserved motifs near the active site, ensuring that catalytic residues remain unaltered after recombination (Sali and Blundell 1993). The use of high-resolution crystal structures or computational models, such as those predicted by AlphaFold, aids in assessing the spatial orientation of active site residues.

(b) Computational Simulation and Interface Design: Molecular dynamics (MD) simulations and molecular docking are invaluable tools for predicting how domain replacement or fusion affects enzyme folding and substrate interactions. MD studies reveal the flexibility and dynamic behavior of active site residues post-recombination, allowing researchers to identify potential distortions or steric conflicts (Karplus and McCammon 2002). Additionally, computational design software such as Rosetta can optimize interdomain interfaces, introducing stabilizing mutations that prevent propagation of structural strain to the active site (Kuhlman and Baker 2000).

(c) Rational Linker Engineering in Fusion Enzymes: In domain fusion, flexible or semi-rigid linkers are often employed to connect catalytic domains without imposing undue stress on the active site. Linker sequences composed of glycine and serine residues provide flexibility, permitting independent folding of each domain. Alternatively, helical or proline-rich linkers can impart rigidity when precise interdomain orientation is required (Amet et al. 2009). Proper linker design minimizes steric hindrance and ensures that substrate channeling between domains is efficient, without compromising the conformation of individual catalytic centers.

(d) Directed Evolution and Site-Directed Mutagenesis: Where rational design cannot fully preserve activity, directed evolution can be employed to restore or enhance the function of the chimeric enzyme. Random mutagenesis followed by high-throughput screening allows the enzyme to "evolve" compensatory mutations that reestablish optimal active site geometry and catalytic efficiency (Arnold 1998). Site-directed mutagenesis can also be used to fine-tune specific residues around the catalytic pocket to improve binding affinity or recover lost activity.

(e) Experimental Characterization and Structural Validation: Once constructed, chimeric enzymes must be thoroughly characterized through kinetic and structural studies. Enzyme kinetics assays reveal whether catalytic efficiency (k_cat/K_m) has been maintained or altered. Structural techniques such as X-ray crystallography, cryo-electron microscopy, and circular dichroism spectroscopy are employed to verify that the overall fold and active site architecture remain intact (Eisenthal and Danson 2002). Comparisons with parent enzyme structures help determine whether any conformational rearrangements have occurred.

A key aspect of maintaining catalytic performance lies in achieving an optimal balance between rigidity, which ensures catalytic precision, and flexibility, which allows substrate accommodation and product release (Tokuriki and Tawfik 2009). Chimeric enzyme design often alters the native dynamic equilibrium of the protein, potentially making the active site too rigid or overly flexible. Maintaining this balance requires preserving essential non-covalent interactions within the active site while allowing dynamic loop regions to retain their mobility.

In some cases, moderate flexibility introduced through domain fusion can actually enhance activity, particularly for enzymes that operate on bulky or diverse substrates. However, excessive flexibility may destabilize the transition state, lowering catalytic efficiency. Therefore, computational simulation and empirical testing are critical for identifying the dynamic range that supports optimal catalytic function.

Several successful examples demonstrate the importance of active site preservation in chimeric enzyme engineering.

Engineering chimeric cellulases by swapping catalytic domains from thermophilic sources has been shown to produce enzymes that largely preserve the parental active-site geometry; this structural conservation can yield enhanced thermostability while maintaining high catalytic efficiency, making domain-swapping a useful strategy for creating robust cellulases for high-temperature applications (García-Paz et al. 2024).

Oxidoreductase Fusions: In fusion constructs combining glucose oxidase and catalase, the individual active sites were carefully preserved through linker optimization, allowing both enzymes to function synergistically in cascade oxidation reactions (Kumar 2020).

Hydrolase Engineering: Domain recombination between esterases and lipases has produced chimeras with broadened substrate ranges while maintaining the catalytic serine-histidine-aspartate triad essential for hydrolysis (Kourist et al. 2010).

Conversely, when active site preservation is neglected—such as through improper domain junctions or mismatched structural frameworks—the resulting chimeric enzymes often lose catalytic activity due to misaligned residues or conformational instability.

Recent advances in computational protein design and machine learning have significantly improved the ability to preserve or redesign active sites during chimeric enzyme construction. Predictive models can now estimate how mutations or recombination events affect catalytic residues and active site electrostatics (Yang et al. 2022). The integration of AlphaFold-based structure prediction with MD simulations enables visualization of atomic-level changes in active site architecture before experimental implementation (Jumper et al. 2021).

Future developments are likely to focus on adaptive active site engineering, where computational algorithms automatically suggest stabilizing mutations that maintain catalytic geometry following domain recombination. The combination of rational design, directed evolution, and synthetic biology will continue to expand the frontiers of chimeric enzyme engineering, allowing scientists to create biocatalysts with both preserved and precisely tuned active sites for industrial, pharmaceutical, and environmental applications.

4.4.3 Scaffold Selection in Chimeric Enzyme Design

Scaffold selection is a crucial and strategic step in the design of chimeric enzymes, as the scaffold serves as the structural backbone that supports the stable integration and optimal functioning of multiple domains. In protein engineering, the term scaffold refers to the core framework or base protein that provides the spatial organization necessary for correct folding, stability, and catalytic performance of the chimeric construct. The scaffold dictates how appended or swapped domains—such as catalytic units, binding motifs, or regulatory elements—are positioned and interact within the overall architecture of the enzyme. Hence, the choice of an appropriate scaffold has a profound impact on the functionality, stability, and expression of the engineered biocatalyst.

An ideal scaffold should possess several key characteristics. It must be structurally robust, capable of withstanding the incorporation of foreign domains without significant loss of folding efficiency or overall stability. Proteins that exhibit high solubility, compact tertiary structure, and tolerance to mutations are often preferred as scaffolds because they minimize the risk of aggregation or misfolding during domain integration. Additionally, a suitable scaffold should have modular flexibility, allowing the accommodation of various domains or linkers without steric hindrance. This ensures that the grafted or fused domains are properly oriented for functional interactions and catalytic cooperation.

Scaffolds are typically selected based on their structural compatibility and evolutionary conservation. Enzymes belonging to the same family often share similar folds and active site architectures, making them promising candidates for domain recombination. For instance, the $(\beta/\alpha)\_8$-barrel, or TIM-barrel fold, is one of the most commonly used scaffolds in chimeric enzyme engineering due to its remarkable stability and tolerance to modifications. This fold has been successfully utilized to host catalytic residues from different enzymes, thereby creating hybrids with altered substrate specificity or enhanced thermal resilience. Similarly, α/β-hydrolase folds, Rossmann folds, and ferredoxin-like folds have also been employed as scaffolds for chimeric enzyme construction.

In addition to structural factors, the biochemical properties of the scaffold play a significant role in determining its suitability. Factors such as pH and temperature tolerance, cofactor dependence, and post-translational modification requirements must align with the intended application of the chimeric enzyme. For example, thermophilic scaffolds are often chosen for industrial biocatalysts designed to function at elevated temperatures, as they confer enhanced robustness and prolonged catalytic lifespan. Conversely, scaffolds derived from mesophilic or psychrophilic organisms are employed when activity under mild or cold conditions is desirable.

From a practical standpoint, scaffold selection also influences expression yield and solubility in heterologous hosts. A scaffold that folds efficiently and expresses at high levels in bacterial or yeast systems simplifies downstream purification and characterization. Engineered scaffolds, such as consensus-designed proteins or de novo-designed frameworks, are increasingly used because they can be tailored for optimal expression and compatibility with diverse domains.

Advances in computational biology have greatly facilitated the process of scaffold selection. Techniques such as homology modeling, molecular dynamics simulations, and machine-learning-based stability prediction enable researchers to evaluate potential scaffolds before experimental implementation. Computational tools can assess interdomain interactions, predict folding outcomes, and identify regions amenable to modification, thereby reducing trial-and-error in laboratory experiments.

In summary, scaffold selection forms the foundation upon which successful chimeric enzyme design is built. A carefully chosen scaffold not only enhances the structural and functional integrity of the engineered enzyme but also ensures that the introduced domains operate synergistically to achieve the desired catalytic performance. As computational and structural biology continue to advance, scaffold design is expected to evolve toward more predictive and rational approaches, paving the way for the creation of highly efficient, stable, and application-specific chimeric enzymes.

4.4.4 Surface Residue Optimization in Chimeric Enzyme Design

Surface residue optimization is a strategic and essential approach in chimeric enzyme design aimed at improving solubility, structural stability, expression efficiency, and resistance to aggregation—while preserving or even enhancing catalytic performance. The amino acid residues exposed on the enzyme's surface govern many crucial interactions, including solvent accessibility, intermolecular associations, and compatibility with the host expression system. Consequently, optimization of these residues enables the fine-tuning of a chimeric enzyme's physicochemical properties, facilitating better folding, higher expression levels, and improved robustness under operational conditions.

In chimeric enzymes—where domains from different proteins are fused or swapped—surface chemistry mismatches can arise due to differences in charge distribution, hydrophobicity, or polarity at the fusion interface. These mismatches often result in misfolding, aggregation, or poor solubility, leading to low yields of functional protein. Surface residue optimization serves as a corrective strategy, ensuring compatibility between domains and maintaining an energetically favorable interface for stable tertiary structure formation (Eijsink et al. 2005). Rational modification of surface residues helps minimize unfavorable interactions and promotes a hydrophilic surface conducive to proper folding and solvation.

A central goal of surface residue optimization is to enhance protein solubility. Aggregation of recombinant proteins, especially in bacterial expression systems like *Escherichia coli*, often stems from excessive surface hydrophobicity or the exposure of non-native hydrophobic patches. Techniques such as surface charge engineering—where positively or negatively charged amino acids (e.g., lysine, arginine, aspartic acid, or glutamic acid) are introduced—can increase electrostatic repulsion between molecules, thereby reducing aggregation propensity (Liu et al. 2019). Similarly, substitution of hydrophobic residues with polar or charged amino acids can render the protein surface more compatible with aqueous environments without disturbing the hydrophobic core that maintains structural integrity (Ladokhin and White 1999).

Another important objective of surface residue optimization is to improve thermostability and conformational resilience. While the enzyme core typically dictates catalytic efficiency, the surface residues play a crucial role in maintaining the global stability of the protein fold under varying conditions of temperature, pH, and ionic strength. Computational approaches such as molecular dynamics simulations, stability prediction algorithms (e.g., FoldX or Rosetta), and machine-learning models are widely used to identify stabilizing mutations at the protein surface (Goldenzweig and Fleishman 2018). These methods allow researchers to predict which amino acid substitutions may enhance the protein's structural rigidity without perturbing the active site or catalytic domain alignment.

Furthermore, surface residue engineering can significantly enhance expression efficiency in heterologous systems. Poorly expressed or insoluble proteins often impose metabolic stress on the host cell, leading to inclusion body formation. Introducing mutations that increase surface polarity or reduce aggregation-prone sequences has been shown to improve soluble expression in systems such as *E. coli, Pichia pastoris*, and insect cell cultures (Sørensen and Mortensen 2005). In the context of chimeric enzymes, surface optimization is particularly valuable when combining domains originating from organisms with different optimal expression environments—for instance, fusing a thermophilic catalytic domain with a mesophilic binding module.

Surface residue modification can also reduce immunogenicity and nonspecific interactions, especially when enzymes are intended for therapeutic or biotechnological applications. Hydrophilic surface engineering and the introduction of consensus sequences derived from non-immunogenic proteins can minimize potential immune recognition (Walsh 2018). Additionally, optimizing residues at interdomain junctions ensures smoother conformational transitions and proper communication between fused domains, which is essential for multi-domain chimeric enzymes performing sequential catalytic reactions.

The optimization process typically follows either a rational design or directed evolution approach. Rational design utilizes computational modeling, structural analysis, and knowledge of amino acid physicochemical properties to propose site-specific mutations. In contrast, directed evolution employs iterative rounds of random mutagenesis and screening to identify beneficial surface variants (Arnold 1998; Romero and Arnold 2009). Combining these strategies—known as semi-rational design—has proven particularly effective in balancing exploration and predictability in enzyme surface optimization (Sun et al. 2019).

In summary, surface residue optimization is a powerful design principle in chimeric enzyme engineering. By strategically modifying surface-exposed amino acids, researchers can overcome challenges associated with domain incompatibility, poor expression, or instability—ultimately achieving chimeric enzymes with superior catalytic, structural, and operational performance. As computational and high-throughput screening technologies continue to advance, surface optimization is poised to become an increasingly precise and predictable tool for next-generation enzyme engineering.

4.4.5 Computational Modeling and Simulation in Chimeric Enzyme Design

Computational modeling and simulation have become indispensable tools in the rational design of chimeric enzymes, providing critical insights into structural compatibility, inter-domain interactions, and dynamic behavior long before experimental validation. The design of chimeric enzymes—proteins that integrate functional elements from distinct sources—requires careful consideration of domain orientation, folding pathways, and active site preservation. Computational approaches enable researchers to visualize, predict, and refine these complex interactions with high precision, significantly reducing the time and cost associated with trial-and-error laboratory experiments.

One of the fundamental applications of computational tools in chimeric enzyme engineering is structure prediction. Modern algorithms such as AlphaFold, Rosetta, and I-TASSER have revolutionized this field by allowing high-resolution, three-dimensional structural modeling based solely on amino acid sequences (Jumper et al. 2021; Yang et al. 2015; Leaver-Fay et al. 2011). These tools predict the spatial arrangement of residues, domain folding, and interfacial contacts, helping assess whether fused domains will adopt stable conformations, preserve catalytic site geometry, and avoid steric clashes. For example, AlphaFold's deep-learning-based predictions have achieved near-experimental accuracy in many cases, enabling reliable modeling of complex chimeric assemblies (Varadi et al. 2022).

In the context of domain swapping and fusion, computational structure prediction serves as a pre-screening mechanism for evaluating multiple fusion designs. Researchers can simulate different linker lengths and attachment points to identify configurations that minimize strain or misalignment at the domain junctions. These models also help maintain active site accessibility and ensure that substrate-binding regions are not obstructed by structural rearrangements following fusion (Eijsink et al. 2005). By comparing alternative structural models, one can rationally select the most promising candidates for experimental expression and testing.

Beyond static structure prediction, molecular dynamics (MD) simulations play a central role in exploring the conformational flexibility and stability of chimeric enzymes over time. MD simulations allow the observation of atomic-level motions, enabling researchers to evaluate inter-domain interactions, folding transitions, and the impact of mutations or linker modifications (Karplus and McCammon 2002;

Hollingsworth and Dror 2018). These simulations can reveal whether domains move cooperatively or independently and identify flexible regions that might cause structural instability. By analyzing trajectories of molecular motion, researchers can estimate properties such as root mean square deviation (RMSD), solvent-accessible surface area, and hydrogen-bond stability—all critical indicators of proper domain integration and enzyme robustness.

Energy minimization and binding free energy calculations further complement these analyses. Using computational frameworks like RosettaDock, AutoDock, or MM/PBSA, scientists can estimate the binding energies between domains or between the enzyme and its substrate, predicting how modifications affect catalytic efficiency (Méndez-Lucio and Medina-Franco 2017). Such methods are particularly valuable when chimeric enzymes are designed for multi-step reactions or when fusion alters substrate channeling between catalytic sites.

Another valuable computational approach is homology modeling, which is employed when experimental structures are unavailable. By aligning the sequence of the designed chimeric enzyme with known structures of homologous proteins, researchers can construct predictive models that guide mutational or structural refinement (Krieger et al. 2009). These models often serve as input for further MD or docking simulations to evaluate structural feasibility and functional potential.

Computational simulations are also crucial in linker design, which is a key aspect of chimeric enzyme construction. The flexibility, length, and amino acid composition of linkers significantly affect domain communication and overall enzyme performance. Simulation-based analyses can determine whether a linker permits sufficient flexibility for substrate transfer while maintaining the relative orientation required for catalytic synergy (Chen et al. 2013). Using coarse-grained or all-atom MD models, researchers can optimize linker sequences to balance rigidity and mobility, ultimately enhancing functional coupling between domains.

In addition to these structure-focused approaches, machine learning (ML) and artificial intelligence (AI) have emerged as transformative tools for predicting enzyme properties and guiding design decisions. ML models trained on large datasets of protein sequences and stability parameters can predict the effects of mutations on folding energy, thermostability, and solubility (Torng and Altman 2019). Integration of ML predictions with MD simulations and structural modeling enables multi-scale optimization, where both global structure and local flexibility are fine-tuned simultaneously.

Overall, computational modeling and simulation represent the cornerstone of modern chimeric enzyme design. These tools bridge the gap between theoretical prediction and experimental realization, allowing for informed decision-making at every design stage—from scaffold selection and domain fusion to active site preservation and stability engineering. As algorithms continue to evolve, combining physics-based simulations with AI-driven prediction models, the rational design of chimeric enzymes is expected to become increasingly accurate, efficient, and predictive, enabling the creation of tailor-made biocatalysts with unprecedented control over structure and function.

4.4.6 Linker Engineering in Chimeric Enzymes

Linker engineering plays a pivotal role in the rational design and functionality of chimeric enzymes, as the linker sequence directly governs structural integrity, conformational flexibility, and cooperative interactions between fused domains. In a chimeric enzyme, distinct protein domains—such as catalytic cores, substrate-binding modules, or regulatory elements—are covalently connected through short peptide linkers that serve as molecular bridges (Chen et al. 2013; Arai et al. 2001). The design of these linkers is crucial, as inappropriate linker composition or length can lead to improper folding, steric clashes, or loss of catalytic activity (George and Heringa 2002).

The type of linker—flexible, rigid, or cleavable—is selected based on the desired spatial and dynamic relationship between domains. Flexible linkers, generally composed of glycine and serine residues (e.g., GGGGS repeats), impart high mobility and minimize steric interference, making them ideal when individual domain independence is required (Chen et al. 2013; Robinson and Sauer 1998). Rigid linkers, which often include proline or charged residues, maintain a defined inter-domain distance and orientation, promoting structural stability and efficient intramolecular communication (Arai et al. 2001). Cleavable linkers, which can be degraded under specific conditions (e.g., protease-sensitive sequences), enable controlled domain separation in multi-functional constructs used in drug delivery or biosensing applications (Wang et al. 2023).

The length of the linker is another key parameter influencing enzyme architecture and functionality. Short linkers can restrict necessary domain motion, potentially causing steric hindrance or misfolding, whereas excessively long linkers can introduce undesirable flexibility that diminishes catalytic efficiency (van Rosmalen et al. 2017). Optimal linker length is typically determined empirically through iterative mutagenesis or computational screening. Recent advances in computational modeling and molecular dynamics simulations—utilizing platforms such as Rosetta, GROMACS, and AlphaFold—have greatly enhanced the capacity to predict domain orientations and assess how linker variations affect stability and dynamics (Baek et al. 2021; Jumper et al. 2021). In certain cases, linkers are designed to form secondary structures, such as α-helices or β-strands, to stabilize inter-domain alignment and minimize entropic penalties (Arai et al. 2001; Wriggers and Schulten 1999).

Experimental validation of linker performance involves heterologous expression of the chimeric construct in systems such as *Escherichia coli* or *yeast*, followed by biochemical characterization of enzyme activity, thermal stability, and substrate specificity (George and Heringa 2002). High-throughput combinatorial approaches—where libraries of linker sequences are systematically screened—have proven effective in identifying optimal configurations (Wang et al. 2023). In industrial biocatalysis, optimized linkers have demonstrated substantial improvements in catalytic efficiency, substrate channeling, and thermal tolerance. For instance, in biofuel production, cellulase chimeras incorporating well-engineered linkers

Table 4.4 Design factors for chimeric enzymes

Design element	Options	Impact on functionality
Domain selection	Catalytic, CBM, redox, regulatory	Determines synergy and expression compatibility
Linker type	Rigid, flexible, protease-sensitive	Affects folding, domain orientation, and activity
Expression host	*E. coli*, yeast, fungi, mammalian cells	Influences glycosylation, folding, and stability

between catalytic and cellulose-binding domains exhibit markedly enhanced cellulose degradation efficiency (Ejaz 2021; Ma et al. 2022). Likewise, in biosynthetic pathways, strategically designed linkers improve multi-enzyme cascade performance by ensuring spatial proximity and efficient substrate transfer (Chen et al. 2013).

In conclusion, linker engineering is not merely a structural consideration but a strategic determinant of chimeric enzyme performance. It bridges computational design, molecular biology, and process engineering to optimize the synergy between functional domains. As predictive modeling and directed evolution tools continue to advance, the rational design of linkers promises to further accelerate the development of efficient, robust, and tailor-made biocatalysts for applications in biotechnology, synthetic biology, and therapeutic innovation.

A summary of considerations for chimeric enzyme design can be found in the Table 4.4.

4.5 Engineering Enzymes to Improve Chimeric Enzyme Functionality

4.5.1 Why Chimeric Enzymes?

Chimeric enzymes are engineered single-chain polypeptides that combine structural and/or catalytic modules from different parental proteins. Their appeal lies in the capacity to:

(1) Co-localize complementary activities (e.g., cellulase and β-glucosidase) to accelerate cascade reactions;

(2) Introduce new substrate recognition motifs or binding domains to increase effective concentration of substrate at catalytic sites;

(3) Engineer multifunctionality for simplified bioprocessing; and

(4) Tune physicochemical tolerance (temperature, pH, organic solvents) by combining domains with different robustness properties.

These advantages have driven both academic studies and industrial interest— especially in lignocellulosic biomass deconstruction, bioremediation, and tailored biosynthesis. Recent literature shows a surge in modular domain-fusion approaches and computational pipelines to predict domain compatibility and linker design (Chaudhari et al. 2023; de María García-Paz et al. 2024; Son et al. 2024).

4.5.2 Principles of Chimeric Enzyme Design

The first design decision when engineering a chimera is which modules to fuse. Typical modules include catalytic domains, substrate-binding carbohydrate-binding modules (CBMs), redox partners, and regulatory domains. Natural modular systems such as cellulosomes and polyketide synthases (PKSs) demonstrate the functional advantages of bringing multiple catalytic units into proximity; these natural paradigms inform synthetic chimeric design by suggesting which domain combinations can yield synergy (Kortemme et al. 2024; Chainani et al. 2025). Domain selection should weigh catalytic compatibility (substrate/product handoff), expression host constraints, and post-translational modifications (glycosylation, disulfide bonds) (Alahuhta et al. 2021; Mitkowski et al. 2024).

A recurrent determinant of chimera performance is the interdomain linker. Linkers mediate relative domain orientation, flexibility, and distance—factors that directly affect substrate channeling and folding. Linker design strategies range from short rigid linkers (to fix orientation), to flexible glycine-rich linkers (to allow independent domain motions), to enzyme-derived linkers that include protease sites or glycosylation motifs. Computational modeling of linker conformational ensembles and experimental screens of linker libraries (including length, composition, and secondary-structure propensity) are widely used to identify linkers that preserve or enhance activity. Recent studies emphasize not only length but also sequence-specific effects on interdomain communication and stability (Jurich et al. 2025; Kortemme et al. 2024).

Fusing domains can create folding interference and aggregation. Strategies to mitigate misfolding include: selecting parental domains known to fold autonomously; placing domains in an order consistent with co-translational folding preferences; engineering glycosylation sites or solubility tags for eukaryotic/extracellular expression; and screening multiple expression hosts (*E. coli*, yeast, filamentous fungi, insect or mammalian cells) to identify systems that provide appropriate chaperones and post-translational modifications. For cellulase chimeras, for example, expression host glycosylation often alters surface charge and stability—factors that have been shown experimentally to affect activity on biomass substrates (Alahuhta et al. 2021; Chaudhari et al. 2023).

4.5.3   Strategies to Improve Functionality

Rational design approaches

Rational design uses structural knowledge to guide modifications: active-site residue substitutions to tune specificity/turnover, interface redesign to improve domain packing, and linker engineering. Structural data (X-ray, cryo-EM) and homology models enable identification of residues that influence substrate access tunnels, electrostatic steering, and domain contacts. Rational design is particularly effective when the mechanism is well understood and the desired change is small (e.g., widening an access channel, removing a steric clash). However, rational approaches struggle with epistasis when multiple residues interact non-additively (Hossack et al. 2023; Kortemme et al. 2024).

Directed evolution and high-throughput screening

Directed evolution (DE) remains a robust route to improve complex, multivariate properties such as catalytic efficiency in non-native contexts or stability under harsh process conditions. For chimeras, DE can be applied at several levels: (a) mutagenize the entire chimera; (b) focus on linkers; or (c) evolve the individual domains prior to fusion. Advances in selection methods (cell-display, microfluidics, droplet screening) and coupled assays (product detection, growth-linked selection) have expanded the feasible search space. Recent work integrates machine learning with DE (active-learning assisted directed evolution) to reduce the number of experimental variants needed to reach optimal solutions (Yang et al. 2025; Hossack et al. 2023).

Computational design and AI-assisted workflows

The last five years have seen a step-change in computational enzyme engineering. Physics-based modeling, molecular dynamics, and AI approaches (deep learning structure prediction and sequence design) offer the ability to evaluate domain compatibility, predict linker conformations, and propose stabilizing mutations. Tools that combine sequence coevolution, structural energetics, and generative models (or hybrid ML/physics pipelines) can propose mutation sets with higher success rates in wet-lab validation than random mutagenesis. In modular chimeras, computational tools can predict unfavorable packing or clashes at interfaces and propose compensatory substitutions, accelerating development cycles (Jurich et al. 2025; Son et al. 2024).

Combining approaches: rational $\rightarrow$ computational $\rightarrow$ DE

A pragmatic engineering workflow often begins with rational design (to limit the search space), applies computational screening to prioritize variants, and finishes with focused directed evolution using high-throughput screens. This integrated paradigm has been successful in producing enzymes with non-natural activity, expanded substrate range, and improved robustness. The combined approach leverages the interpretability of rational design, the scale of computation, and the empirical power of evolution (Hossack et al. 2023; Yang et al. 2025).

4.5.4 Representative Case Studies

Chimeric cellulases for lignocellulosic biomass deconstruction

Cellulolytic systems are classic targets for chimeric engineering because biomass deconstruction benefits from close proximity of multiple enzyme activities and from strong substrate binding domains. Examples include chimeric cellobiohydrolases and multifunctional cellulase constructs where catalytic domains (endoglucanase, exoglucanase, β-glucosidase) are fused with CBMs or dockerin/cohesin modules to mimic cellulosome architectures. Experimental results demonstrate that chimeric cellulases can increase product yields and modify processivity, but outcomes depend heavily on domain order, linker length/composition, and host expression glycosylation patterns. One experimental study expressed a chimeric cellobiohydrolase I in several oleaginous yeasts to compare glycosylation and activity; differences in stability and biomass digestion highlighted the interplay between host processing and chimera performance (Alahuhta et al. 2021; Mitkowski et al. 2024).

In order to optimize the design of chimeric enzymes, one should test a range of domain arrangements to ensure correct folding and activity. Moreover, multiple linker sequences for both flexibility and stability should be evaluated. It is important to choose expression hosts with care, considering their glycosylation profiles and secretion pathways to maximize yield and functionality.

Engineered laccases and lignin-targeting chimeras

Laccases are multi-copper oxidases widely studied for lignin modification, dye decolorization, and biosensor applications. Engineering efforts have produced chimeric laccases fused with accessory domains (e.g., carbohydrate-binding motifs or peroxidase partners) to increase substrate scope and processivity. Recent engineered laccases from fungal sources have shown improved catalytic efficiency on lignin model compounds and enhanced stability under industrially relevant conditions (Pham LTM et al. 2024; Guan et al. 2025). Such chimeras are promising for lignin valorization in biorefineries.

Multifunctional xylanases and hemicellulases (trifunctional chimeras)

Beyond cellulases, chimeric fusion of xylanase with accessory esterase or feruloyl esterase modules yields multifunctional enzymes that can act in a concerted manner on complex hemicellulose structures. Reports of engineered trifunctional xylanases demonstrate improved substrate conversion in model and real lignocellulosic substrates, though linker design and positional effects (N- vs. C-terminal fusions) strongly influence activity. These chimeras reduce the need for enzyme cocktails, simplifying downstream processing (de María García-Paz et al. 2024).

Chimeric Hydrogenases and Bioenergy Enzymes

Chimeric approaches have been explored for hydrogenases and other energy-relevant enzymes to improve electron transfer, oxygen tolerance, and catalytic turnover. Some studies leverage fusion to redox partners or protective catalytic partners (e.g., catalase) to shield sensitive active sites from reactive oxygen species, thereby improving turnovers in whole-cell or cell-free systems. Directed evolution campaigns targeted at chimeric hydrogenases have reported improved hydrogen production in engineered strains, illustrating the feasibility of combining modular design with evolutionary optimization for bioenergy applications (Plummer et al. 2016).

4.5.5   *Experimental Workflows and Practical Considerations*

A DBTL cycle for chimeric enzyme engineering commonly includes:

(1) **Design**—select domains, design linkers, and propose mutations (rational/computational);
(2) **Build**—construct genetic assemblies using modular cloning (Golden Gate, Gibson, or DNA assembly pipelines) and express in multiple hosts if needed;
(3) **Test**—use high-throughput assays (colorimetric, fluorescence, growth coupling, mass spectrometry) to quantify both activity and stability;
(4) **Learn**—analyze sequence-function relationships with statistical or ML models to guide the next cycle. Active-learning integration can notably reduce the number of variants screened by selecting experiments that maximize information gain (Yang et al. 2025; Hossack et al. 2023).

Selecting the right assay is crucial. For cascade chimeras, assays that measure the final product (rather than intermediate activity) better capture the functional advantages of co-localization. For industrial relevance, tests under process-mimicking conditions (high solids loading, presence of inhibitors, non-neutral pH, elevated temperature) provide realistic performance metrics. Coupled enzymatic assays and product detection by chromatography or mass spectrometry can identify bottlenecks in substrate handoff between fused domains (Chaudhari et al. 2023; Son et al. 2024).

Scale-Up and Process Integration

A chimeric enzyme that performs in microtiter assays may still fail at pilot scale due to viscosity, product inhibition, or proteolytic degradation. Considerations for scale-up include thermo- and protease-stability engineering, immobilization (to recycle biocatalyst), and compatibility with downstream separation. Chimeras designed for immobilization can incorporate binding modules or tags to enable oriented attachment that preserves interdomain geometry (Mitkowski et al. 2024; ASM Journals).

4.5.6 Challenges and Limitations

Interdomain epistasis—where mutations or domain arrangements interact non-additively—complicates prediction. Small changes in linker sequence or domain order can lead to large unpredictable shifts in activity or stability. This makes exhaustive exploration of sequence/arrangement space infeasible for large systems and underscores the value of integrated computational and evolutionary strategies (Kortemme et al. 2024; Hossack et al. 2023).

Fusion proteins can be susceptible to host proteases, particularly at linker junctions. Engineering protease-resistant linkers, choosing secretory pathways with lower protease activity, or co-expressing protease inhibitors are practical routes to mitigate degradation. Additionally, glycosylation differences between host species can alter folding and surface properties, which can be either beneficial or deleterious (Alahuhta et al. 2021; Mitkowski et al. 2024).

Trade-Offs Between Specificity and Promiscuity

Fusing domains to create multifunctional chimeras may broaden substrate scope but risk reducing catalytic efficiency on any particular substrate. Engineering must balance desired specificity with the practical gains from multifunctionality (e.g., fewer enzymes in a cocktail vs. peak per-enzyme activity). Directed evolution can sometimes restore high activity while retaining multifunctionality, but this often requires careful screening design (Yang et al. 2025; Hossack et al. 2023).

4.5.7 Future Directions

Emerging physics-based and ML-hybrid pipelines show promise in evaluating domain interfaces and predicting linker ensembles that favor functional coupling. As structure prediction (e.g., deep learning) and MD sampling improve, in silico filters will more reliably eliminate incompatible designs before experimental testing. This reduces experimental burden and accelerates timelines for chimeric development (Jurich et al. 2025; Son et al. 2024).

Automated retrobiosynthesis and modular assembly frameworks that propose chimeric PKS and enzyme fusions for complex product synthesis are emerging. These pipelines—combining retrosynthetic logic with domain fusion design—enable creation of novel metabolic routes with fewer genetic parts and tighter substrate channeling. Continued progress will lower the barrier to designing bespoke biosynthetic factories (Chainani et al. 2025; Kortemme et al. 2024).

Combining active-learning algorithms with extremely high-throughput screening (droplet microfluidics, cell-free expression coupled to mass spectrometry readouts) will enable exploration of larger design spaces for chimeras. These methods prioritize variants that maximally improve model predictions, making DE campaigns more efficient for complex traits like multifunctional activity (Yang et al. 2025; Son et al. 2024).

The long-term vision is libraries of orthogonal domains and standardized linkers that reliably assemble into functional chimeras across hosts and processes. Standardization (akin to synthetic biology parts registries) combined with robust design rules will enable predictable engineering of multifunctional biocatalysts for industrial deployment (de María García-Paz et al. 2024; Mitkowski et al. 2024).

Chimeric enzymes represent a powerful strategy to engineer multifunctional biocatalysts with advantages for industrial bioprocessing, biosynthesis, and environmental applications. The field has matured from simple domain fusions to integrated engineering pipelines that combine rational design, computational prediction, and directed evolution. Representative successes in cellulase, laccase, xylanase, and hydrogenase systems illustrate both the potential and the barriers—most notably epistasis, host effects, and linker sensitivity. Ongoing advances in structural computation, machine learning, and high-throughput directed evolution promise to reduce uncertainty and accelerate practical deployment of chimeric enzymes in the coming years (Alahuhta et al. 2021; Chaudhari et al. 2023; Yang et al. 2025).

4.6 Potential Applications of Chimeric Enzymes in Pulp and Paper Industry

4.6.1 *Limitations of Conventional Chemical Treatments*

The pulp and paper industry has historically relied on intensive chemical treatments to achieve key processing objectives such as delignification, bleaching, and deinking. These operations commonly employ reagents like chlorine, chlorine dioxide, sodium hydroxide, and other strong alkalis to break down lignin, remove ink particles, and purify cellulose fibers. While these chemical methods are effective in achieving desired pulp brightness and cleanliness, they come with significant environmental and economic drawbacks. The effluents generated from such treatments often contain high concentrations of toxic compounds, including chlorinated organics and adsorbable organic halides (AOX), which pose serious risks to aquatic ecosystems and human health. As a result, mills are required to invest heavily in advanced wastewater treatment infrastructure to comply with environmental regulations, further increasing operational costs (Bajpai 2010).

Moreover, the procurement, storage, handling, and disposal of these hazardous chemicals introduce additional layers of complexity and risk. Workers are exposed to potential chemical burns, inhalation hazards, and accidental spills, necessitating stringent safety protocols and personal protective equipment. The volatility of chemical prices and supply chain disruptions can also impact production continuity and profitability. In light of these challenges, there has been a growing interest in developing alternative technologies that are both environmentally sustainable and economically viable.

4.6.2 Emergence of Enzymatic Alternatives

Enzymatic treatments have emerged as a promising solution to address these concerns. Enzymes offer a biologically driven approach to fiber modification and contaminant removal, operating under milder conditions and generating fewer harmful byproducts (Bajpai 2018a, b). Their specificity and catalytic efficiency make them attractive candidates for replacing or supplementing traditional chemical processes. However, native enzymes—those derived directly from microbial or fungal sources—often fall short in industrial settings. They typically exhibit narrow operational ranges in terms of temperature and pH, limited substrate affinity, and reduced stability under the harsh conditions prevalent in pulp mills (Pathak et al. 2017; Yakubo et al. 2019; Yang et al. 2023). These limitations hinder their scalability and consistent performance in continuous processing environments (Bajpai 1999a, b, 2018a,b).

4.6.3 Chimeric Enzymes as Next-Generation Biocatalysts

To overcome these constraints, chimeric enzymes have gained increasing attention as next-generation biocatalysts tailored for industrial applications. These engineered proteins are constructed by combining catalytic and binding domains from different biological sources, resulting in multifunctional molecules with enhanced activity, stability, and substrate versatility. Chimeric enzymes are being actively explored to improve enzymatic efficiency, extend operational tolerance across a broader range of temperatures and pH levels, and enable synergistic catalytic activity within complex pulp matrices. Their modular design allows for rational customization to meet specific process requirements, such as selective lignin degradation, improved fiber flexibility, enhanced ink detachment, or pitch control (García-Paz et al. 2024; Martins et al. 2020; Gilmore et al. 2020; Jadaun et al. 2020; Dahiya et al. 2024; Shahid et al. 2023; Chen et al. 2019; Arsalan et al. 2025; Lee et al. 2011; Yang et al. 2016).

Structurally, chimeric enzymes integrate multiple catalytic or binding modules into a single polypeptide chain, enabling them to perform multiple functions simultaneously or sequentially. This integration can take various forms: for instance, fusing catalytic domains—such as combining cellulase with lipase to simultaneously loosen fibers and hydrolyze resinous pitch deposits—can streamline processing steps and reduce enzyme dosing complexity. Alternatively, attaching carbohydrate-binding modules (CBMs) to catalytic cores can significantly enhance substrate affinity and localization, improving hydrolytic efficiency on insoluble fiber surfaces.

Another strategy involves co-expressing multiple lignin-degrading enzymes, such as laccase and peroxidase, within a single microbial host to facilitate oxidative delignification in biobleaching applications.

4.6.4 Technical Design Considerations

The successful design and deployment of chimeric enzymes for industrial pulp and paper applications demand meticulous attention to several interdependent technical and operational factors. At the forefront is the necessity to ensure robust enzymatic activity and structural integrity under the alkaline pH conditions that prevail in kraft pulping environments, typically ranging from pH 8.5 to 11.0. Enzymes must retain catalytic efficiency and avoid denaturation or deactivation in these chemically aggressive settings. Equally important is the preservation of proper protein folding and domain orientation, especially when multiple catalytic or binding modules are fused into a single polypeptide chain. Misfolding or steric clashes between domains can severely compromise enzyme functionality, reduce substrate accessibility, and increase susceptibility to aggregation or proteolytic degradation.

Compatibility with existing industrial infrastructure is another critical design consideration. Chimeric enzymes must be tailored to integrate seamlessly with current process flow parameters, including temperature profiles, residence times, mixing regimes, and fiber consistency levels. Their activity should align with the throughput and operational constraints of continuous digesters, bleaching towers, and deinking units. Enzymes that require specialized equipment or non-standard conditions may face barriers to adoption due to retrofitting costs and process disruptions.

4.6.5 Expression Systems and Production Platforms

In addition to functional design, the choice of expression system plays a pivotal role in determining the commercial viability of chimeric enzymes. The selected microbial host must support high-yield production, efficient secretion into the extracellular medium, and minimal downstream purification complexity. Commonly used hosts include *Escherichia coli*, which offers rapid growth and ease of genetic manipulation but may lack post-translational modifications; *Pichia pastoris*, a methylotrophic yeast capable of glycosylation and high-density fermentation; and *Bacillus subtilis*, a Gram-positive bacterium known for its secretion efficiency and Generally Recognized As Safe (GRAS) status. Each system presents distinct advantages and trade-offs in terms of scalability, regulatory acceptance, and process economics.

Moreover, the expression construct must be optimized for codon usage, promoter strength, signal peptide selection, and plasmid stability to ensure consistent and high-level production. Fermentation parameters such as pH, aeration, and nutrient composition must be fine-tuned to maximize yield and minimize proteolytic degradation. The downstream processing pipeline—including filtration, concentration, and formulation—should be designed to preserve enzyme activity and stability during storage and transport.

4.6.6   Industrial Benefits of Chimeric Enzymes

When these design principles are effectively implemented, chimeric enzymes can deliver transformative benefits to the pulp and paper industry. They enable significant reductions in chemical dependency, thereby lowering environmental impact and effluent toxicity. Enhanced enzymatic specificity and multifunctionality contribute to improved fiber modification, increased pulp brightness, and superior paper strength. Additionally, their integration into existing workflows can reduce energy consumption, simplify process control, and support compliance with evolving environmental regulations. As such, chimeric enzymes represent a strategic innovation pathway for advancing sustainable and high-performance bioprocessing in the pulp and paper sector.

4.6.7   Enzymatic Deconstruction of Lignocellulosic Biomass

Breaking down the polymeric components of plant cell walls presents a major biochemical challenge due to the intricate, cross-linked, and chemically diverse composition of lignocellulosic biomass. Plant cell walls primarily consist of cellulose, hemicellulose, and lignin—three biopolymers that are tightly interwoven through covalent and non-covalent interactions, forming a highly recalcitrant matrix. To deconstruct this complex architecture, a synergistic system of enzymes is required. These include:

- Endoglucanases and exoglucanases (which cleave internal and terminal β-1,4-glycosidic bonds in cellulose, respectively),
- Xylanases and β-xylosidases (which degrade hemicellulose components such as xylan),
- Laccases (which oxidatively modify lignin).

The cooperative activity of these enzymes is essential for achieving efficient hydrolysis of lignocellulosic substrates and generating fermentable sugars for biotechnological applications, including biofuel production and biopulping.

In nature, some anaerobic microorganisms, particularly species of *Clostridium* and *Ruminococcus*, have evolved specialized enzyme complexes known as cellulosomes, which represent one of the most efficient systems for lignocellulose deconstruction. Cellulosomes are multienzyme assemblies anchored to scaffoldin proteins that spatially organize cellulases, hemicellulases, and accessory enzymes for coordinated hydrolysis. Inspired by this natural model, researchers have engineered artificial cellulosome-like systems and multifunctional fusion proteins by integrating cellulosomal domains with oxidative or hydrolytic enzymes such as laccases.

Davidi et al. (2016) engineered a dockerin-fused laccase (Tfu1114) and incorporated it into a designer cellulosome together with cellulases and a xylanase. Inclusion of the laccase increased delignification and roughly doubled the amount of reducing sugars released from wheat straw versus the same enzyme complex without the laccase, demonstrating that coupling lignin-oxidizing activity to scaffolded enzyme complexes improves cellulase access to cellulose microfibrils.

4.6.8   Case Studies of Chimeric Enzyme Applications

Another promising strategy to improve biomass degradation involves the construction of chimeric proteins, which integrate distinct catalytic or binding domains into a single polypeptide chain. These artificial enzymes often combine complementary catalytic functions, such as those of cellulases and xylanases, thereby enabling the simultaneous breakdown of multiple polysaccharide components. Researchers have successfully applied the domain insertion technique to produce multifunctional chimeras. For example, the lipase–phospholipase A1 chimera, also referred to as lecitase ultra, demonstrates superior degumming performance at both acidic and elevated temperature conditions. Whereas the native lipase and phospholipase exhibit optimal activity at pH 3.5–4.2, the chimeric enzyme remains active at pH 5.5 and retains substantial catalytic efficiency at 60 °C, illustrating how chimeric design can broaden enzyme stability and operational flexibility (Khamies et al. 2024).

A notable and well-characterized example of chimeric enzyme application in pulp and paper processing is the Lipase-Endoglucanase chimera (Lip-EG1CD), which was successfully expressed in *Pichia pastoris*. This engineered construct combines the catalytic domains of lipase and endoglucanase, enabling simultaneous hydrolysis of hydrophobic toner residues and cellulose fibers. In enzymatic deinking trials, Lip-EG1CD achieved approximately 89% toner removal, demonstrating a significant improvement over conventional enzyme blends. Additionally, the treated pulp exhibited enhanced optical properties, including increased brightness, attributes critical for recycled paper quality (Liu et al. 2017). Handsheet strength also showed a notable improvement, underscoring the effectiveness of the enzymatic intervention. The findings indicate that the combined deinking action of endoglucanase and lipase on wastepaper substrates can be significantly enhanced through the construction of an appropriately designed chimeric enzyme. Such a chimera promotes intramolecular synergistic interactions between the two catalytic domains, thereby increasing overall efficiency compared to the use of individual enzymes applied separately. This enhanced synergistic performance presents a promising pathway for developing a more economical and scalable strategy for wastepaper recycling. In addition, the fusion-based approach minimizes reliance on harsh surfactants and reduces the extent of mechanical agitation typically required in conventional deinking processes. These reductions translate into improved energy efficiency, lower operational costs, and a more environmentally sustainable recycling workflow. Ultimately, the integration of such chimeric biocatalysts can support greener industrial practices while maintaining or even improving the quality of recycled pulp.

Recent developments have extended this concept to the degradation of complex lignocellulosic matrices. A laccase–glucanase chimera created through domain insertion improved sugar release from milled sugarcane bagasse by ~ 20% compared with an equimolar mixture of the separate enzymes (Furtado et al. 2013).

Guo et al. (2013) indicated that the engineered tandem fusions of endo-xylanase and endo-mannanase exhibited improved efficiency in the depolymerization of hemicellulose within biomass substrates.

A xylosidase–arabinofuranosidase chimera, designed via gene fusion, increased xylose yield ($\approx$30%) from wheat arabinoxylan compared with the separate enzyme mixture (Fan et al. 2009a, b, c).

Another innovative construct—a laccase–endoxylanase chimera assembled by insertional fusion—demonstrated improved catalytic turnover, enhanced thermo-tolerance and greater resistance to proteolysis while maintaining both oxidative and hydrolytic activity, features advantageous for harsh industrial pulp & paper environments (Ribeiro et al. 2011).

4.6.9 Advances in Thermostability and Catalytic Efficiency

Recent progress in chimeric enzyme engineering has focused on improving thermostability and catalytic turnover for industrial bioprocesses. For example, fusion constructs that combine xylanase and β-xylosidase domains have produced chimeric enzymes that release more xylose from wheat arabinoxylan (~30% increase in one case) and exhibit altered pH/temperature behavior and enhanced thermal resilience of the xylosidase domain (Fan et al. 2009a, b, c).

Such improvements in thermal tolerance and enzymatic efficiency are indispensable for sustaining continuous industrial operations, particularly in sectors such as pulp and paper, biofuels, and food processing, where enzymes are routinely exposed to extreme pH values and high-temperature environments over extended durations. The integration of thermostable domains within chimeric constructs not only enhances operational robustness but also reduces enzyme deactivation rates, thereby improving process economics and reliability.

The successful design, construction, and heterologous expression of chimeric cellulases in *Escherichia coli*, as reported by Du et al. (2021), provide compelling evidence for the efficacy of molecular fusion strategies in enhancing enzymatic performance. These recombinant enzymes were engineered by appending cellulose-binding modules (CBMs) to the catalytic core, resulting in a structurally integrated biocatalyst with superior substrate interaction capabilities. Quantitative assays revealed a substantial 58% increase in hydrolytic efficiency when tested against recalcitrant substrates such as Avicel and filter paper. This enhancement was primarily attributed to the improved substrate affinity conferred by the CBMs, which function as non-catalytic auxiliary domains. By anchoring the catalytic region directly onto the cellulose surface, CBMs facilitate a localized increase in enzyme concentration at the substrate interface, thereby improving substrate accessibility and turnover rates under industrially relevant conditions.

In a parallel study, Carli et al. (2022) demonstrated the functional advantages of modular domain engineering by integrating a synthetic carbohydrate-binding module with an endogalacturonase enzyme. This chimeric construct exhibited up to a 35% improvement in catalytic efficiency during lignocellulosic biomass saccharification. The synthetic CBM enhanced the enzyme's ability to interact with complex plant polysaccharides, thereby accelerating the breakdown of pectic components and improving overall saccharification yields. These findings underscore the strategic value of domain fusion in tailoring enzyme architectures for specific industrial applications, including biofuel production, pulp and paper processing, and agricultural waste valorization. Collectively, these studies validate the modular engineering approach as a robust platform for developing next-generation biocatalysts with enhanced functionality, stability, and process compatibility.

4.6.10 Regulatory and Transcriptional Chimeras

Beyond catalytic fusions, researchers are also designing regulatory chimeras to optimize enzyme expression networks. A novel transcription factor, GaaR–XlnR, was recently identified as a bifunctional activator and xylanolytic regulator that induces pectinolytic enzyme synthesis in *Aspergillus niger* when exposed to D-xylose. Such regulatory chimeras provide a new level of control in fungal enzyme systems and can be harnessed to fine-tune industrial fermentation processes.

A study by Kun et al. (2021) reported the discovery of a novel chimeric transcription factor, GaaR–XlnR, which functions as both a transcriptional activator and a regulator of xylanolytic gene expression. This engineered regulator enhances the synthesis of key pectinolytic enzymes in *Aspergillus niger*, particularly when d-xylose is present as an inducer. By integrating functional domains from two distinct native regulators, the GaaR–XlnR chimera demonstrates improved control over polysaccharide-degrading pathways, offering valuable insights into the fine-tuning of fungal metabolism for biotechnological applications. In the pulp-and-paper industry, similar fusion-based strategies are being effectively employed. For instance, a gene encoding a chimeric enzyme comprising esterase (GDEst-95) and lipase domains has been successfully utilized to enhance enzymatic performance in fiber processing and pitch control. Such multifunctional biocatalysts streamline industrial workflows by combining complementary activities within a single polypeptide chain, thereby improving catalytic efficiency while reducing enzyme dosage and overall processing costs. Collectively, these advancements highlight the growing potential of chimeric regulatory proteins and fusion enzymes in optimizing industrial bioprocesses, particularly those requiring coordinated degradation of complex plant-derived polymers.

4.6.11 Industrial Bio-Bleaching Applications

A chimeric laccase–xylanase enzyme developed for bio-bleaching applications demonstrated markedly superior performance compared to its individual parental enzymes. In experimental evaluations, the fusion enzyme exhibited approximately 1.5-fold higher catalytic activity, highlighting the beneficial synergistic effects arising from the combination of oxidative and hemicellulolytic functions within a single polypeptide. Thermostability assessments further underscored the robustness of the chimeric construct. After 2 h of incubation at 60 °C, the laccase–xylanase chimera retained nearly 70% of its initial activity, whereas the native laccase and xylanase enzymes preserved only 40% and 45%, respectively, under the same conditions (Saadat 2017). This enhanced stability is particularly valuable for industrial bio-bleaching processes, which often involve elevated temperatures and extended reaction times. The improved catalytic efficiency and thermal resilience of the chimeric enzyme translate into practical advantages such as reduced enzyme loading, shorter processing times, and more effective lignin modification. Collectively, these attributes position laccase–xylanase chimeras as promising candidates for developing energy-efficient, cost-effective, and environmentally sustainable bleaching strategies in the pulp and paper industry.

Chimeric enzyme formed by fusing β-1,4-glucosidase and GH5 endoglucanase (F194A mutant) displayed a higher Tm at 78 °C, a decrease in Km (0.02 mM pNPG) implying an increase in specificity and an increase in the catalytic efficiency (Nath et al. 2019).

In another investigation, researchers engineered six distinct chimeric enzymes by creating various domain combinations of naturally occurring xylanase–cellulase from *Clostridium thermocellum* and β-glucosidase from *C. cellulovorans*. These constructs were generated by fusing the catalytic domains at either the N-terminal or C-terminal ends, allowing systematic evaluation of how domain orientation influences overall enzyme performance. The resulting chimeric enzymes demonstrated significantly enhanced functional properties. Notably, several variants displayed improved thermal and pH stability, indicating greater resilience under the harsh conditions typically encountered in biomass processing. More importantly, the optimized constructs showed a 2.3-fold increase in catalytic efficiency toward carboxymethyl cellulose (CMC), leading to more effective production of amorphous cellulose, a key intermediate in lignocellulosic biomass deconstruction (Chen et al. 2019). These findings highlight the potential of rational domain fusion to tailor multifunctional enzymes with superior catalytic behavior. By integrating complementary glycoside hydrolase activities into single polypeptide chains, such chimeric biocatalysts can streamline biomass conversion, reduce enzyme dosage requirements, and improve the overall efficiency of industrial saccharification processes.

A growing body of empirical evidence (see Table 4.5) strongly supports the conclusion that chimeric enzymes consistently surpass their native counterparts in key performance metrics, including catalytic efficiency, substrate specificity, and structural stability under operational conditions. These enhancements are particularly evident in engineered xylanases, which have been extensively modified through advanced protein engineering techniques such as domain insertion and end-to-end ligation. These strategies enable the rational design of multifunctional enzyme constructs tailored to optimize the hydrolysis of complex hemicellulosic substrates.

Table 4.5 Chimeric enzymes exhibiting possible uses in the pulp and paper sector

eric enzyme	Enzyme source	Plasmid	Expression Host	Methodology	Operational advantage	Linker	Reference
Mannanase-Xylanase	Acetivibrio thermocellus	pET28a	E. coli BL21 (DE3)	End to end fusion	Improvement of the thermal stability, kinetic parameters (kcat), and storage stability of the enzyme	–	(Ariaeenejad et al. 2021)
Xylanase-Feruloyl esterase	Prevotella ruminicola	pANY2	E. coli BL21 (DE3)	End to end fusion	Improved hydrolysis of wheat straw and sugarcane bagasse with commercial cellulase	–	(Wang et al. 2022)
Xylanase-β-Glucosidase	Cattle rumen metagenomic data	pET28a	E. coli BL21 (DE3)	End to end fusion	Improved degradation efficiency of organic solvent pretreated coffee residue waste. Activity improvement after immobilisation is fourfold at 80 °C and up to twofold at pH 4.0	–	(Ariaeenejad et al. 2023)

(continued)

Table 4.5 (continued)

eric enzyme	Enzyme source	Plasmid	Expression Host	Methodology	Operational advantage	Linker	Reference
Mannanase-Xylanase	Bacillus subtilis	pETSn	E. coli BL21 (DE3)	Post-translational conjugation	Enhanced thermostability and ion stability; resistance to agglomeration and freeze–thaw treatment	–	(Gao et al. 2019)
Lipase-Cutinase	Pichia pastoris	pPICZ αA	E. coli DH5α	End to end fusion	Twofold higher activity compared to individual enzymes; improved sheet brightness	–	(Liu et al. 2017)
Endo-1,4-β-Xylanase-β-Xylosidase	Bacillus subtilis	pRARE2	E. coli BL21 (DE3)	Overlap extension PCR	Threefold increase in xylose production and enhanced thermal stability	Tpet 0899 ETITGEVDDCEYEQAQQQAG PEVTYE	(Gilmore et al. 2020)
Xylanase-Arabinofuranosidase	Clostridium thermocellum	pET29b	E. coli BL21 (DE3)	End to end fusion	Enhanced pH and thermal stability; enhanced catalytic activity compared to parent enzyme	GGGGGADQLAIGPMYNQVVYQYPN	(Fan et al. 2009a, b, c)

(continued)

Table 4.5 (continued)

eric enzyme	Enzyme source	Plasmid	Expression Host	Methodology	Operational advantage	Linker	Reference
Xylanase-Xylosidase	Clostridium thermocellum	pET29b	E. coli BL21 (DE3)	End to end fusion	Enhanced pH and thermal stability; enhanced catalytic activity compared to parent enzyme	GGGGGADQLAIGPMYNQVVYQYPN	(Fan et al. 2009a, b, c)
Feruloyl esterase A-Endoxylanase B	Escherichia coli JM109	pGEM-T	Aspergillus niger	Overlap extension PCR	Increased substrate selectivity and higher production of ferulic acid	GSTYSSGSSSGSGSSSSS	(Wang et al. 2020)
Xylanase–Lichenase	Bacillus subtilis	pET	E. coli BL21 (DE3)	Overlap extension PCR	Similar activity as parent enzymes with enhanced substrate specificity	4-glycine linker	(Lin et al. 2018)
Xylanase-Mannanase	Escherichia coli TOP10F'	pGAPZαA / pUM-T	Pichia pastoris	Overlap extension PCR	Improved hydrolysis of luffa cylindrical fibre	rigid α-helix-forming peptide A [EAAAK]3 A	(Guo et al. 2013)

(continued)

Table 4.5 (continued)

eric enzyme	Enzyme source	Plasmid	Expression Host	Methodology	Operational advantage	Linker	Reference
Xylanase-Acetylxylan esterase	Neocallimastix patriciarum	pET-29a	E. coli BL21 (DE3)	Overlap extension PCR	Greater hydrolytic activity on natural xylans and rice straw than parental enzymes with minor operational adjustments	(GGGGS)2	(Martins et al. 2020)
Lipase-Endoglucanase	E. coli DH5α	pPICZ αA	Pichia pastoris	Domain insertion	Improved ink removal efficiency and increased brightness; enhanced sheet strength	–	(Liu et al. 2017)
Xylanase–Cellulase	Gloeophyllum trabeum	pPICZ A	Pichia pastoris GS115	Overlap extension PCR	1.4-fold increase in hydrolysis activity and high yield production of fermentable sugar	(GGGGS)2	(Kim et al. 2015)
Endoglucanase (Endo5A)-Xyl11D	Paenibacillus ICGEB2008 strain	pUC18 / pQE30	E. coli DH5α	End to end fusion	More than twofold activity enhancement compared to individual enzymes	glycine-serine linker (GGGGSGGGGS)	(Adlakha et al. 2011)

(continued)

Table 4.5 (continued)

eric enzyme	Enzyme source	Plasmid	Expression Host	Methodology	Operational advantage	Linker	Reference
Mannanase–Xylanase	Bacillus subtilis	pETSn	E. coli BL21 (DE3)	Post-translational conjugation	Improved thermostability, ion stability, resistance to agglomeration, and freeze–thaw tolerance	–	(Gao et al. 2019)
XYN3–XYN4 Endoxylanase	Aspergillus niger	pET29b	E. coli BL21 (DE3)	Post-translational conjugation	Improved xylan hydrolysis (twofold increase) and nearly threefold increase in catalytic efficiency	–	(Saadat 2017)

Arsalan et al. (2025). Reproduced with permission

A representative study by Jadaun et al. (2020) exemplifies this approach, wherein four distinct chimeric constructs were developed to combine xylanase and mannanase activities within a single polypeptide framework. The domains were linked using a synthetic A[EAAAK]$_3$ flexible linker, which facilitates spatial separation and conformational flexibility between catalytic regions, thereby preserving individual domain functionality. These constructs were successfully expressed in the eukaryotic host *Pichia pastoris* X33, known for its robust secretion capabilities and post-translational modification machinery. Among the engineered variants, the construct designated XE3M demonstrated the most pronounced improvement in enzymatic performance, particularly in the degradation of *Luffa cylindrica* fibers. Quantitative saccharification assays revealed significantly elevated sugar release rates, indicating enhanced catalytic synergy and substrate accessibility.

The ability to efficiently hydrolyze structurally complex plant polymers into fermentable sugars holds substantial industrial relevance. In the context of biopulping, such enzymatic systems can reduce chemical input and energy consumption while improving fiber quality. Moreover, the liberated sugars serve as critical feedstocks for the synthesis of bio-based fuels, platform chemicals, and value-added bioproducts. This dual utility underscores the strategic importance of chimeric enzyme platforms in advancing integrated biorefinery models and promoting the transition toward a circular bioeconomy, as emphasized by Saadat (2017). The modularity and tunability of these constructs offer a versatile framework for future innovations in sustainable biotechnology.

The fusion of xylanase with carbohydrate-binding modules (CBMs) derived from *Nonomuraea flexuosa* has shown promising results in enhancing enzymatic performance. The CBMs facilitate stronger and more targeted interactions with hemicellulosic substrates, thereby increasing hydrolytic efficiency without promoting undesirable lignin adsorption. This selective action is particularly important in prebleaching applications, where excessive lignin binding can interfere with fiber integrity and downstream brightness development. The engineered xylanase-CBM construct demonstrated improved substrate affinity and catalytic turnover, making it a viable candidate for integration into enzymatic prebleaching workflows (Zhang et al. 2013).

Another compelling example involves recombinant *Bacillus subtilis* engineered to co-express laccase and lignin peroxidase enzymes. These oxidative biocatalysts play a pivotal role in lignin degradation, and their simultaneous expression within a single host streamlines the biobleaching process. The multi-enzyme system exhibited superior delignification efficiency compared to single-enzyme treatments, resulting in higher pulp brightness and reduced chemical demand. This approach also minimizes enzyme dosing complexity and enhances process control, offering a scalable alternative to chlorine-based bleaching agents (Ozer et al. 2018). Collectively, these examples underscore the potential of chimeric and co-expressed enzyme systems to improve pulp quality, reduce environmental impact, and optimize operational economics in modern paper manufacturing.

Pycnoporus cinnabarinus laccase was fused to the C-terminal linker and carbohydrate binding module (CBM) of *Aspergillus niger* cellobiohydrolase B (CBHB). The chimeric enzyme of molecular mass 100 kDa was successfully produced in *A. niger*. Laccase-CBM was further purified to determine its main biochemical properties. The Michaelis–Menten constant and pH activity profile were not modified, but the chimeric enzyme was less thermostable than either the *P. cinnabarinus* laccase or the recombinant laccase produced in the same strain. Laccase-CBM was able to bind to a cellulosic substrate and, to a greater extent, to softwood kraft pulp. Binding to the pulp was shown to be mainly time and temperature-dependent. Laccase-CBM was further investigated for its softwood kraft pulp biobleaching potential and compared with the *P. cinnabarinus* laccase. Addition of a CBM was shown to greatly improve the delignification capabilities of the laccase in the presence of 1-hydroxybenzotriazole (HBT). In addition, ClO(2) reduction using 5 U of chimeric enzyme per gram of pulp was almost double than that observed using 20 U of *P. cinnabarinus* laccase per gram of pulp. It has been demonstrated that conferring a carbohydrate binding capability to the laccase could significantly enhance its biobleaching properties (Ravalason et al. 2009).

The multifunctionality of chimeric enzymes translates into several industrial benefits: reduced chemical consumption, enhanced pulp brightness, improved mechanical properties of paper, and lower toxicity in effluents. Their ability to perform multiple catalytic functions simultaneously simplifies process workflows, reduces energy input, and minimizes the need for sequential enzyme dosing or chemical pretreatment (Landwehr et al. 2007; Sakhuja et al. 2021). This integration of functions into a single molecule also facilitates better control over reaction kinetics and process reproducibility.

4.6.12 *Challenges and Limitations*

Despite their considerable promise in improving process efficiency and sustainability, chimeric enzymes continue to face a range of technical and economic hurdles that limit their widespread adoption in industrial pulp and paper operations. One of the foremost challenges is protein folding instability, which arises when multiple functional domains—often with distinct structural requirements—are fused into a single polypeptide chain. Improper folding can lead to misalignment of active sites, reduced catalytic efficiency, and aggregation, ultimately compromising enzyme performance. Additionally, these engineered proteins are frequently susceptible to proteolytic degradation, especially under the harsh chemical and thermal conditions prevalent in pulp mills. This degradation not only reduces enzyme longevity but also necessitates higher dosing, thereby increasing operational costs.

Another major barrier is the high cost of production, particularly when scaling up from laboratory expression systems to industrial-scale fermentation. Producing chimeric enzymes in microbial hosts such as *Escherichia coli*, *Pichia pastoris*, or *Bacillus subtilis* often requires complex optimization of codon usage, promoter strength, and secretion pathways. Downstream purification and stabilization further add to the expense, making cost-effectiveness a critical concern for commercial viability. Moreover, achieving a balanced catalytic profile—where each domain contributes effectively without interfering with others—is a delicate task. For example, excessive cellulase activity within a chimeric construct can lead to over-degradation of cellulose fibers, resulting in fiber shortening, loss of tensile strength, and diminished paper quality (Yadav et al. 2023). Fine-tuning enzyme kinetics to preserve fiber integrity while maximizing process benefits remains a key area of ongoing research.

Beyond technical constraints, regulatory and biosafety considerations present additional layers of complexity. The use of genetically modified organisms (GMOs) for enzyme production requires rigorous documentation, including strain lineage, genetic modifications, containment protocols, and environmental risk assessments. Regulatory approval processes vary across jurisdictions and often involve lengthy evaluations by agencies such as the EPA, EFSA, or national biosafety committees. Compliance with these standards is essential not only for legal authorization but also for public acceptance and market access. Furthermore, intellectual property protection for novel chimeric constructs must be carefully managed to safeguard innovation while enabling collaborative development.

Recent advances in directed evolution, computational protein modeling, and thermostable enzyme engineering are significantly accelerating the development of robust chimeric constructs tailored for industrial applications. Directed evolution enables iterative refinement of enzyme properties through high-throughput mutagenesis and selection, while computational modeling facilitates rational design by predicting domain compatibility, folding dynamics, and active site geometry. Thermostable enzyme engineering, in particular, has proven critical for enhancing operational resilience under the elevated temperatures and alkaline conditions typical of pulp and paper processing environments. These synergistic innovations are collectively transforming enzyme design from empirical trial-and-error approaches to predictive, structure-guided methodologies.

As a result, the research paradigm is undergoing a marked shift away from conventional enzyme cocktails—where multiple enzymes are dosed separately—toward the creation of single-molecule multifunctional chimeras. These engineered proteins integrate multiple catalytic or binding domains into a unified scaffold, offering streamlined biocatalysis with reduced formulation complexity. Such chimeras are increasingly recognized for their potential to deliver cost-effective, scalable, and environmentally sustainable solutions for industrial bioprocessing, particularly in sectors like pulp and paper, textile, and biofuels (Planas-Iglesias et al. 2021a, b; Chainani et al. 2025). Their modularity also allows for customization to specific substrate profiles and process conditions, enhancing both efficiency and adaptability.

However, translating these promising laboratory constructs into full-scale industrial operations remains a formidable challenge. It requires comprehensive techno-economic analysis to evaluate cost–benefit ratios, enzyme production scalability, and integration with existing infrastructure. Pilot-scale demonstrations are essential to validate enzyme performance under realistic flow rates, substrate loads, and process variability. Moreover, long-term performance monitoring is critical to assess enzyme stability, resistance to proteolytic degradation, and consistency across production batches. Without these validation steps, even the most promising chimeric designs may fail to meet the rigorous demands of commercial deployment.

Technical challenges and limitations of chimeric enzymes are presented in Table 4.6.

Despite extensive research on native and recombinant enzymes for pulp and paper bioprocessing—particularly xylanases, cellulases, and laccases—there remains a notable paucity of peer-reviewed studies specifically addressing the design, characterization, and industrial deployment of chimeric enzymes within this sector. While chimeric constructs have demonstrated enhanced catalytic efficiency, thermostability, and substrate specificity in broader lignocellulosic biomass conversion contexts (e.g., biofuels, agricultural waste valorization), their targeted application in pulp bleaching, fiber modification, deinking, and pitch control has not been systematically explored or validated.

Current literature predominantly focuses on single-domain enzymes or recombinant variants expressed in microbial hosts, with limited integration of modular fusion strategies such as domain ligation, linker optimization, or carbohydrate-binding module (CBM) incorporation tailored to pulp substrates. Moreover, there is a lack of comparative studies evaluating the performance of chimeric constructs against native enzymes under industrial pH, temperature, and chemical load conditions typical of kraft, sulfite, and mechanical pulping processes.

This gap underscores the need for interdisciplinary investigations combining enzyme engineering, process modeling, and pilot-scale validation to assess the feasibility and regulatory compatibility of chimeric biocatalysts in pulp and paper manufacturing. Addressing this gap could significantly advance sustainable processing, reduce chemical inputs, and support circular bioeconomy initiatives.

The deployment of chimeric enzymes in the pulp and paper industry presents both exciting opportunities and important regulatory challenges. On one hand, multifunctional chimeras promise to streamline operations by consolidating catalytic activities—such as bleach boosting, pitch control, and deinking—into single biocatalysts, thereby reducing chemical usage, energy demand, and effluent toxicity. This aligns strongly with sustainability goals and emerging green chemistry standards. On the other hand, regulatory frameworks will require rigorous validation of these engineered constructs, including assessments of safety, allergenicity, environmental persistence, and process compliance to ensure that no unintended by-products compromise pulp quality or effluent safety. Looking ahead, pilot-scale trials of xylanase–laccase hybrids, lipase–esterase chimeras, and cellulase–LPMO fusions will be critical in demonstrating both efficacy and compliance. If successful, these innovations could redefine enzymatic processing in pulp and paper, shifting from complex enzyme cocktails to tailored, multifunctional biocatalysts that meet industrial performance targets while satisfying stringent environmental and regulatory standards (Table 4.6).

Table 4.6 Technical challenges and limitations of chimeric enzymes

Challenge area	Description	Industrial impact	Reference/Examples
Protein folding instability	Fusion of multiple domains can cause misfolding, steric clashes, or aggregation	Reduced catalytic efficiency, loss of activity, increased susceptibility to degradation	Yadav et al. (2023)
Proteolytic degradation	Harsh mill conditions accelerate enzyme breakdown	Shortened enzyme lifespan, higher dosing requirements, increased costs	General pulp mill conditions
Over-degradation of fibers	Excess cellulase activity may shorten fibers and reduce tensile strength	Lower paper quality, diminished mechanical properties	Yadav et al. (2023)
Production costs	Complex optimization of codon usage, promoters, secretion pathways; costly downstream purification	Limits scalability and commercial viability	Du et al. (2021); Carli et al. (2022)
Domain balance	Difficulty in achieving balanced catalytic profiles across fused domains	Risk of one activity dominating, reducing synergistic benefits	Chen et al. (2019)
Thermostability limits	Some chimeras lose activity under extended high-temperature exposure	Reduced robustness in continuous operations	Ravalason et al. (2009)
Regulatory and biosafety	GMO-based enzyme production requires extensive documentation and risk assessment	Slows adoption, adds compliance burden	Kun et al. (2021); Regulatory frameworks

4.7 Forward-Looking Perspectives on Chimeric Enzymes for Sustainable Paper Production

4.7.1 Market Outlook for Enzyme Technologies

The pulp and paper industry, long regarded as one of the most resource-intensive sectors, is undergoing a profound transformation driven by sustainability imperatives, regulatory pressures, and technological innovation. Among the most promising developments is the application of "chimeric enzymes", engineered biomolecules that combine functional domains from different enzymes to achieve enhanced catalytic efficiency, substrate specificity, and operational stability. Forward-looking perspectives on their role in pulp and paper production highlight not only their potential to revolutionize conventional processes but also their ability to align industrial practices with global sustainability goals.

From a market perspective, the adoption of advanced enzyme technologies in the pulp and paper sector is expected to accelerate significantly over the coming decade. The global pulp and paper enzymes market, which was valued at **USD 92.5 million in 2024**, is projected to reach **USD 125.4 million by 2032**, reflecting a steady compound annual growth rate (CAGR) of **3.7%**. This trajectory underscores the increasing recognition of enzymatic treatments as superior alternatives to conventional chemical-intensive methods, offering both operational efficiency and environmental benefits that resonate strongly with industry stakeholders and regulators alike (https://chemicalresearchinsight.com/2025/10/17/pulp-and-paper-enzymes-market-global-outlook-and-forecast-2025–2032/). As the industry faces mounting challenges related to energy consumption, chemical usage, and waste generation, chimeric enzymes offer a pathway toward greener, more efficient, and economically viable operations.

4.7.2 Future Perspectives

At present, enzymes such as xylanases, cellulases, and laccases are already widely used in pulp and paper manufacturing for applications including prebleaching, deinking, pitch control, and fiber modification. However, natural enzymes often suffer from limitations such as narrow pH and temperature ranges, susceptibility to inhibitors, and reduced activity in industrial conditions. Chimeric enzymes, by contrast, are designed to overcome these constraints by fusing catalytic domains or binding modules from different enzymes into a single construct. This structural innovation enables them to perform multiple functions simultaneously, tolerate harsh environments, and maintain activity across diverse substrates. For example, a chimeric enzyme combining xylanase and cellulase domains could simultaneously degrade hemicellulose and cellulose residues, thereby improving pulp fibrillation and reducing the need for mechanical refining. Such multifunctionality is particularly valuable in integrated mills where process efficiency and cost reduction are paramount (Arsalan et al. 2024).

Looking ahead, one of the most significant forward-looking perspectives is the role of chimeric enzymes in "reducing chemical dependency" in pulp bleaching. Traditional bleaching relies heavily on chlorine-based or chlorine dioxide chemicals, which generate toxic effluents and contribute to environmental pollution. Enzyme-assisted bleaching, particularly with xylanases, has already demonstrated reductions in chlorine consumption (Bajpai 1999a, b, 2006, 2009, 2013, 2018a,b; Bajpai and Bajpai 1997; Viikari et al. 1986; Suurnakki et al. 1997; Singh et al. 2016; Kuhad et al. 1997; Bhat 2000)) Chimeric enzymes could amplify this effect by combining oxidative and hydrolytic activities, enabling deeper lignin removal and improved fiber brightness with minimal chemical input. This not only reduces environmental impact but also lowers operational costs associated with effluent treatment and chemical procurement (Arsalan et al. 2025). In the long term, the integration of chimeric enzymes into bleaching sequences could pave the way for fully enzyme-driven biobleaching processes, eliminating the need for hazardous chemicals altogether.

Another forward-looking perspective involves the application of chimeric enzymes in "fiber recycling and deinking" With the global emphasis on circular economy principles, the demand for recycled paper is increasing. However, recycling processes face challenges such as ink removal, fiber damage, and loss of bonding strength. Chimeric enzymes designed to combine lipase, cellulase, and hemicellulase activities could simultaneously break down ink binders, loosen fiber surfaces, and restore fiber bonding capacity. This multifunctional approach would enhance the quality of recycled pulp, reduce the need for harsh alkaline chemicals, and extend the life cycle of fibers. Moreover, enzyme-assisted recycling could help mills meet regulatory targets for waste reduction and resource efficiency, positioning the industry as a leader in sustainable manufacturing.

The potential of chimeric enzymes also extends to "energy efficiency". Mechanical refining, a critical step in papermaking, consumes significant amounts of energy to modify fiber morphology and improve bonding. By deploying chimeric enzymes that combine fibrillation-enhancing activities with water retention properties, mills could reduce refining intensity and energy consumption. This aligns with global efforts to decarbonize industrial operations and reduce greenhouse gas emissions. Furthermore, the ability of chimeric enzymes to operate under mild conditions reduces the need for high-temperature processes, further contributing to energy savings. In the long run, enzyme-driven fiber modification could replace mechanical refining altogether, representing a paradigm shift in papermaking technology.

From a regulatory perspective, the adoption of chimeric enzymes supports compliance with increasingly stringent environmental standards. Governments and international bodies are imposing tighter restrictions on effluent discharge, chemical usage, and carbon emissions. Mills that integrate chimeric enzymes into their processes will be better positioned to meet these requirements while maintaining competitiveness. Additionally, enzyme-based processes generate fewer secondary pollutants, simplifying effluent treatment and reducing regulatory risks. This forward-looking alignment between biotechnology and regulation underscores the strategic importance of chimeric enzymes in ensuring the long-term viability of the pulp and paper industry (Arsalan et al. 2025).

The development of chimeric enzymes also opens new avenues for "product innovation" Beyond traditional paper, the industry is diversifying into specialty products such as nanocellulose, bio-based packaging, and functional papers with enhanced properties. Chimeric enzymes could facilitate the production of nanocellulose by combining cellulase and oxidative domains, enabling efficient breakdown of fibers into nanoscale components. Similarly, enzymes with tailored binding modules could modify fiber surfaces to impart hydrophobicity, antimicrobial activity, or improved printability. These innovations not only expand market opportunities but also align with consumer demand for sustainable and high-performance materials. In this sense, chimeric enzymes are not merely process enhancers but also enablers of product diversification and value creation.

A forward-looking perspective must also consider the "economic implications" of enzyme adoption. While enzyme production and engineering involve upfront costs, the long-term savings in chemicals, energy, and effluent treatment often outweigh these expenses. Advances in synthetic biology and fermentation technology are expected to reduce enzyme production costs, making chimeric enzymes more accessible to mills of varying scales. Furthermore, partnerships between biotechnology firms and pulp and paper companies could accelerate commercialization, ensuring that enzyme innovations are tailored to industrial needs. The economic viability of chimeric enzymes will be further strengthened by carbon credits, green certifications, and consumer preference for eco-labeled products, all of which provide financial incentives for sustainable practices.

In terms of research and development, the future of chimeric enzymes lies in "precision engineering and computational design", advances in protein modeling, machine learning, and directed evolution are enabling scientists to predict enzyme structures, optimize domain combinations, and enhance stability. This computational approach reduces trial-and-error experimentation and accelerates the development of industrially relevant enzymes. Moreover, the integration of omics technologies—genomics, proteomics, and metabolomics—provides insights into microbial diversity and enzyme functionality, expanding the pool of candidates for chimeric design. As these tools mature, the pace of innovation in enzyme engineering will increase, bringing forth a new generation of highly efficient biocatalysts tailored for pulp and paper applications.

Another forward-looking dimension is the "integration of chimeric enzymes with digital technologies" Smart manufacturing systems, driven by sensors and artificial intelligence, could monitor enzyme activity in real time, adjusting dosages and process parameters for optimal performance. This synergy between biotechnology and digitalization enhances process control, reduces variability, and maximizes enzyme efficiency. In the future, mills could operate as fully integrated biorefineries, where chimeric enzymes and digital systems work in tandem to optimize resource utilization and minimize environmental impact. Such integration represents the convergence of Industry 4.0 and green chemistry, positioning the pulp and paper sector at the forefront of sustainable industrial innovation.

Despite these promising perspectives, challenges remain. The stability of chimeric enzymes under industrial conditions must be rigorously validated, and large-scale production must be economically feasible. Intellectual property considerations, regulatory approvals, and public acceptance of genetically engineered enzymes also require careful management. Nevertheless, the trajectory of enzyme technology in pulp and paper suggests that these challenges are surmountable. Collaborative efforts among academia, industry, and regulatory bodies will be essential to ensure safe, effective, and widespread adoption of chimeric enzymes. Future Insights on Chimeric Enzymes are presented in Table 4.7.

Table 4.7 Future insights on chimeric enzymes

Application area	Role of chimeric Enzymes	Industrial benefit
Market outlook	Engineered multifunctional enzymes projected to expand market share	Growth of global enzyme market (USD 92.5 M in 2024 → USD 125.4 M by 2032; CAGR 3.7%)
Bleaching	Fusion of oxidative + hydrolytic domains for deeper lignin removal	Reduced chlorine use, lower effluent toxicity, cost savings in wastewater treatment
Fiber recycling & deinking	Lipase–cellulase–hemicellulase chimeras for ink removal and fiber restoration	Higher pulp brightness, stronger recycled fibers, reduced alkaline chemical demand
Energy Efficiency	Enzymes enhancing fibrillation and water retention	Lower refining intensity, reduced energy consumption, alignment with decarbonization goals
Regulatory compliance	Enzyme-driven processes replacing hazardous chemicals	Easier compliance with effluent, chemical, and carbon emission standards
Product innovation	Tailored chimeras for nanocellulose, bio-packaging, functional papers	Diversification into high-value sustainable products
Economic implications	Synthetic biology and fermentation lowering production costs	Long-term savings, carbon credits, eco-label incentives
Precision engineering	Computational design, machine learning, omics integration	Faster development of stable, efficient industrial enzymes
Digital integration	AI-driven monitoring of enzyme activity in smart mills	Real-time optimization, reduced variability, Industry 4.0 synergy
Challenges	Stability validation, large-scale production, IP/regulatory hurdles	Need for collaborative R&D and regulatory frameworks

In conclusion, forward-looking perspectives on chimeric enzymes in the pulp and paper industry reveal a transformative potential that extends across environmental, economic, and technological dimensions. By reducing chemical dependency, enhancing recycling, improving energy efficiency, and enabling product innovation, chimeric enzymes address the core challenges facing the industry. Their development is supported by advances in synthetic biology, computational design, and digital integration, ensuring that future enzymes are more efficient, stable, and versatile. While challenges remain, the alignment of chimeric enzyme technology with sustainability goals and regulatory frameworks positions it as a cornerstone of the industry's future. As mills transition toward greener and smarter operations, chimeric enzymes will play a pivotal role in shaping a sustainable and competitive pulp and paper sector for decades to come.

References

Adlakha N, Rajagopal R, Kumar S, Reddy VS, Yazdani SS (2011) Synthesis and characterization of chimeric proteins based on Cellulase and xylanase from an insect gut bacterium. Appl Environ Microbiol 77:4859–4866. https://doi.org/10.1128/AEM.02808-10

Alahuhta M, Xu Q, Knoshaug EP, Wang W, Wei H, Amore AD, Baker JO, Vander Wall TA, Himmel ME, Zhang M (2021). Chimeric cellobiohydrolase I expression, activity, and biochemical properties in three oleaginous yeast. Biotechnol Biofuels 14(6). https://doi.org/10.1186/s13068-020-01856-z

Al-Khayri JM, Abdel Latef AAH, Taha HSA, Eldomiaty AS, Abd-Elfattah MA, Rezk AA, Shehata WF, Almaghasla MI, Shalaby TA, Sattar MN, Awad MF, Hassanin AA (2022) In silico profiling of proline biosynthesis and degradation related genes during fruit development of tomato. SABRAO J Breed Genet 54:549–564. https://doi.org/10.54910/sabrao2022.54.3.8

Amet N, Lee HF, Shen WC (2009) Insertion of the designed helical linker led to increased stability of a recombinant fusion protein. Pharm Res 26(3):523–528

Anwar Z, Gulfraz M, Irshad M (2014) Agro-industrial lignocellulosic biomass a key to unlock the future bio-energy: a brief review. J Radiat Res Appl Sci 7(2):163–173. https://doi.org/10.1016/j.jrras.2014.02.003

Arai R, Ueda H, Kitayama A, Kamiya N, Nagamune T (2001) Design of the linkers which effectively separate domains of a bifunctional fusion protein. Protein Eng 14(8):529–532. https://doi.org/10.1093/protein/14.8.529

Ariaeenejad S, Motamedi E, Kavousi K, Ghasemitabesh R, Goudarzi R, Salekdeh GH, Zolfaghari B, Roy S (2023) Enhancing the ethanol production by exploiting a novel metagenomic-derived bifunctional xylanase/β-glucosidase enzyme with improved β-glucosidase activity by a nanocellulose carrier. Front Microbiol 13. https://doi.org/10.3389/fmicb.2022.1056364

Arnold FH (1998) Design by directed evolution. Acc Chem Res 31(3):125–131

Arnold FH (2018) Directed evolution: bringing new chemistry to life. Angew Chem Int Ed 57(16):4143–4148. https://doi.org/10.1002/anie.201708408

Arsalan A, Ravikumar Y, Tang X, Cao Z, Zhao M, Sun W, Qi X (2025) Chimeric enzymes in the pulp and paper making industry: current developments. Biotechnol Adv 79:108530. https://doi.org/10.1016/j.biotechadv.2025.108530

Aÿ J, Götz F, Borriss R, Heinemann U (1998) Structure and function of the *Bacillus* hybrid enzyme GluXyn-1: native-like jellyroll fold preserved after insertion of autonomous globular domain. Proc Natl Acad Sci U S A 95:6613–6618

Baek M, DiMaio F, Anishchenko I, Dauparas J, Ovchinnikov S, Lee GR, Wang J, Cong Q, Kinch LN, Schaeffer RD, Millán C, Park H, Adams C, Glassman CR, DeGiovanni A, Pereira JH, Rodrigues AV, van Dijk AA, Ebrecht AC, Opperman DJ, Sagmeister T, Buhlheller C, Pavkov-Keller T, Rathinaswamy MK, Dalwadi U, Yip CK, Burke JE, Garcia KC, Grishin NV, Adams PD, Read RJ, Baker D (2021) Accurate prediction of protein structures and interactions using a three-track neural network. Science 373(6557):871–876. https://doi.org/10.1126/science.abj8754. Epub 2021 Jul 15. PMID: 34282049; PMCID: PMC7612213

Bajpai P (1999a) Application of enzymes in the pulp and paper industry. Biotechnol Progr 15:147–157. https://doi.org/10.1016/j.jnoncrysol.2003.08.068

Bajpai P (2010) Environmentally friendly production of pulp and paper. Wiley VCH

Bajpai P (1999b) Application of enzymes in the pulp and paper industry. Biotechnol Prog 15(2):147–157

Bajpai P (2006) Potential of biotechnology for energy conservation in pulp and paper. In: Energy management for pulp and papermakers; Oct 16–18, Paper 11; Budapest, Hungary, pp 29

Bajpai P (2009) Xylanases. In: Schaechter M, Lederberg J (eds) Encyclopedia of microbiology, vol 4, 3rd edn. Academic Press, San Diego, CA, pp 600–612

Bajpai P (2013) Pulp and paper bioprocessing. Encycl Ind Biotechnol 1–17

Bajpai P (2018a) Biotechnology for pulp and paper processing. Springer Nature, Singapore

Bajpai P (2018b) Industrial enzymes—an update, first ed. Bookboon, UK, pp 118. ISBN:978-87-403-2129-6

Bajpai P, Bajpai PK (1997) Realities and trends in enzymatic prebleaching of kraft pulp. In: Scheper T (ed) Advances in biochemical engineering and biotechnology, vol 56. Springer-Verlag, Berlin, Germany, pp 1–31

Baldwin RL (1999) Structure and mechanism in protein science. A guide to enzyme catalysis and protein folding, by A. Fersht. 1999. New York: Freeman. 631 pp. $67.95 (hardcover). Protein Science 2000;9(1):207–207. https://doi.org/10.1110/ps.9.1.207

Bashirova A, Pramanik S, Volkov P, Rozhkova A, Nemashkalov V, Zorov I, Gusakov A, Sinitsyn A, Schwaneberg U, Davari MD (2019) Disulfide bond engineering of an endoglucanase from *Penicillium verruculosum* to improve its Thermostability. Int J Mol Sci 20:1602. https://doi.org/10.3390/ijms20071602

Bashton M, Chothia C (2007) The generation of new protein functions by the combination of domains. Structure 15(1):85–99

Beck A, Reichert JM (2011) Therapeutic Fc-fusion proteins and peptides as successful alternatives to antibodies. Mabs 3(5):415–416. https://doi.org/10.4161/mabs.3.5.17334

Betran E, Long M (2003) Dntf-2r, a young drosophila retroposed gene with specific male expression under positive darwinian selection. Genetics 164(3):977–988

Bhat MK (2000) Cellulases and related enzymes in biotechnology. Biotechnol Adv 18(5):355–83. https://doi.org/10.1016/s0734-9750(00)00041-0. PMID: 14538100

Bornscheuer UT, Pohl M (2001) Improved biocatalysts by directed evolution and rational protein design. Curr Opin Chem Biol 5(2):137–143

Bornscheuer UT, Huisman GW, Kazlauskas RJ, Lutz S, Moore JC, Robins K (2012) Engineering the third wave of biocatalysis. Nature 485(7397):185–194

Bulow L, Mosbach K (1996) Fusion proteins. Ann N Y Acad Sci 799(1):376–382

Carli S, Parras Meleiro L, Salgado JCS, et al (2022) Synthetic carbohydrate-binding module-endogalacturonase chimeras increase catalytic efficiency and saccharification of lignocellulose residues. Biomass Conv Bioref 14:6369–6380 (2024). https://doi.org/10.1007/s13399-022-027 16-6

Chainani Y, Diaz J, Guilarte-Silva M, Blay V, Zhang Q, Sprague W, Tyo KEJ, Broadbelt LJ, Mukhopadhyay A, Keasling JD, Martin HG, Backman TWH (2025) Merging the computational design of chimeric type I polyketide synthases with enzymatic pathways for chemical biosynthesis. Nat Commun 16(1):5787. https://doi.org/10.1038/s41467-025-61160-y

Chaudhari YB, Várnai A, Sørlie M, Horn SJ, Eijsink VGH (2023) Engineering cellulases for conversion of lignocellulosic biomass. Protein Eng Des Sel 36:gzad002. https://doi.org/10.1093/protein/gzad002

Chen Z, Wang L, Shen Y, Hu D, Zhou L, Lu F, Li M (2022) Improving thermostability of chimeric enzymes generated by domain shuffling between two different original glucoamylases. Front Bioeng Biotechnol. 10:881421. https://doi.org/10.3389/fbioe.2022.881421

Chen C-H, Yao J-Y, Yang B, Lee H-L, Yuan S-F, Hsieh H-Y, Liang P-H (2019) Engineer multi-functional cellulase/xylanase/β-glucosidase with improved efficacy to degrade rice straw. Bioresour Technol Rep 5:170–177. https://doi.org/10.1016/j.biteb.2019.01.008

Chen X, Zaro JL, Shen WC (2013) Fusion protein linkers: property, design and functionality. Adv Drug Deliv Rev 65(10):1357–1369. https://doi.org/10.1016/j.addr.2012.09.039

Chung HS, Louis JM, Eaton WA (2009) Experimental determination of upper bound for transition path times in protein folding from single-molecule photon-by-photon trajectories. Proc Natl Acad Sci 106:11837–11844

Crasto CJ, Feng J (2000) LINKER: a program to generate linker sequences for fusion proteins. Protein Eng des Sel 13(5):309–312. https://doi.org/10.1093/protein/13.5.309

Crum MA, Park JM, Sewell BT, Benedik MJ (2015) C-terminal hybrid mutant of *Bacillus pumilus* cyanide dihydratase dramatically enhances thermal stability and pH tolerance by reinforcing oligomerization. J Appl Microbiol 118(4):881–889

Cui Y, Cui W, Liu Z, Zhou L, Kobayashi M, Zhou Z (2014) Improvement of stability of nitrile hydratase via protein fragment swapping. Biochem Biophys Res Commun 450(1):401–408

Czajkowsky DM, Hu J, Shao Z, Pleass RJ (2012) Fc-fusion proteins: new developments and future perspectives. EMBO Mol Med 4(10):1015–1028. https://doi.org/10.1002/emmm.201201379

Dahiya S, Rapoport A, Singh B (2024) Biotechnological potential of lignocellulosic biomass as substrates for fungal xylanases and its bioconversion into useful products: a review. Fermentation 10:82. https://doi.org/10.3390/fermentation10020082

Dan WB, Zhang C, Guan ZB, Zhang SQ (2008) The construction of bifunctional fusion proteins consisting of duck BAFF and EGFP. Biotechnol Lett 30(2):221–227. https://doi.org/10.1007/s10529-007-9543-y

Davidi L, Moraïs S, Artzi L, Knop D, Hadar Y, Arfi Y, Bayer EA (2016) Toward combined delignification and saccharification of wheat straw by a laccase-containing designer cellulosome. Proc Natl Acad Sci U S A 113(39):10854–10859. https://doi.org/10.1073/pnas.1608012113

Diogo JA, Hoffmam ZB, Zanphorlin LM, Cota J, Machado CB, Wolf LD, Squina F, Damásio AR, Murakami MT, Ruller R (2015) Development of a chimeric hemicellulase to enhance the xylose production and thermotolerance. Enzyme Microb Technol 69:31–37. https://doi.org/10.1016/j.enzmictec.2014.11.006

Doolittle RF (1995) The multiplicity of domains in proteins. Annu Rev Biochem 64:287–314

dos Santos AM, da Costa CHS, Martins M, Goldbeck R, Skaf MS (2024) Exploring the structural and dynamic properties of a chimeric glycoside hydrolase protein in the presence of calcium ions. Int J Mol Sci 25:11961. https://doi.org/10.3390/ijms252211961

Du L, Cui X, Li H, Wang Y, Fan L, He R, Jiang F, Yu A, Xiao D, Ma L (2021) Enhancing the enzymatic hydrolysis efficiency of lignocellulose assisted by artificial fusion enzyme of swollenin-xylanase. Ind Crop Prod 173:114106. https://doi.org/10.1016/j.indcrop.2021.114106

Eijsink VG, Gåseidnes S, Borchert TV, van den Burg B (2005) Directed evolution of enzyme stability. Biomol Eng 22(1–3):21–30. https://doi.org/10.1016/j.bioeng.2004.12.003. PMID: 15857780

Eisenthal R, Michael JD (eds) Enzyme assays: a practical approach (Oxford, 1992; online edn, Oxford Academic, 31 Oct. 2023), https://doi.org/10.1093/oso/9780199638208.001.0001. Accessed 12 Nov. 2025

Ejaz U, Sohail M, Ghanemi A (2021) Cellulases: from bioactivity to a variety of industrial applications. Biomimetics (Basel) 6(3):44. https://doi.org/10.3390/biomimetics6030044. PMID: 34287227; PMCID: PMC8293267

Elleuche S (2015) Bringing functions together with fusion enzymes—from nature's inventions to biotechnological applications. Appl Microbiol Biotechnol 00(2015):1545–1556. https://doi.org/10.1007/s00253-014-6315-1

Fan Z, Wagschal K, Lee CC, Kong Q, Shen KA, Maiti IB, Yuan L (2009a) The construction and characterization of two xylan-degrading chimeric enzymes. Biotechnol Bioeng 102(3):684–692. https://doi.org/10.1002/bit.22112

Fan Z, Werkman JR, Yuan L (2009b) Engineering of a multifunctional hemicellulase. Biotechnol Lett 31(5):751–757. https://doi.org/10.1007/s10529-009-9926-3

Fan Z, Yuan L (2010) Production of multifunctional chimaeric enzymes in plants: a promising approach for degrading plant cell wall from within. Plant Biotechnol J 8(3):308–315. https://doi.org/10.1111/j.1467-7652.2009.00484.x

Fan Z, Wagschal K, Chen W, Montross MD, Lee CC, Yuan L (2009c) Multimeric hemicellulases facilitate biomass conversion. Appl Environ Microbiol 75:1754–1757

Ferreira P, Fernandes PA, Ramos MJ (2022) Modern computational methods for rational enzyme engineering. Chem Catal 2:2481–2498. https://doi.org/10.1016/j.checat.2022.09.036

Furtado GP, Ribeiro LF, Lourenzoni MR, Ward RJ (2013) A designed bifunctional laccase/beta-1,3–1,4-glucanase enzyme shows synergistic sugar release from milled sugarcane bagasse. Protein Eng Des Sel. 26(1):15–23

Furtado GP, Santos CR, Cordeiro RL, Ribeiro LF, de Moraes LAB, Damasio ARL, Polizeli M de LTM, Lourenzoni MR, Murakami MT, Ward RJ (2015) Enhanced xyloglucan-specific endo-β-1,4-glucanase efficiency in an engineered CBM44-XegA chimera. Appl Microbiol Biotechnol 99:5095–5107. https://doi.org/10.1007/s00253-014-6324-0

Gao D-Y, Sun X-B, Liu M-Q, Liu Y-N, Zhang H-E, Shi X-L, Li Y-N, Wang J-K, Yin S-J, Wang Q, (2019) Characterization of thermostable and chimeric enzymes via Isopeptide bond-mediated molecular cyclization. J Agric Food Chem 67:6837–6846. https://doi.org/10.1021/acs.jafc.9b01459

García-Paz FM, Del Moral S, Morales-Arrieta S, Ayala M, Treviño-Quintanilla LG, Olvera-Carranza C (2024) Multidomain chimeric enzymes as a promising alternative for biocatalysts improvement: a minireview. Mol Biol Rep 51(1):410. https://doi.org/10.1007/s11033-024-09332-9

George RA, Heringa J (2002) An analysis of protein domain linkers: their classification and role in protein folding. Protein Eng 15(11):871–879. https://doi.org/10.1093/protein/15.11.871

Gilbert M, Bayer R, Cunningham AM, DeFrees S, Gao Y, Watson DC, Young NM, Wakarchuk WW (1998) The synthesis of sialylated oligosaccharides using a CMP-Neu5Ac synthetase/sialyltransferase fusion. Nat Biotechnol 16(8):769–772. https://doi.org/10.1038/nbt0898-769

Gilmore SP, Lillington SP, Haitjema CH, de Groot R, O'Malley MA (2020) Designing chimeric enzymes inspired by fungal cellulosomes. Synth Syst Biotechnol 5(1):23–32

Goldenzweig A, Fleishman SJ (2018) Principles of protein stability and their application in computational design. Annu Rev Biochem 87:105–129. https://doi.org/10.1146/annurev-biochem-062917-012102

Graveley (2001) Alternative splicing: increasing diversity in the proteomic world. Trends Genet 17(2):100–107

Guan Z, Yao W, Ye J, Dan W, Shen J, Zhang S (2007) The construction and characterization of a bifunctional EGFP/sAPRIL fusion protein. Appl Microbiol Biotechnol 73:1114–1122

Guan G, Li B, Xu L, Qian J, Zou B, Huo S, Ding Z, Cui K, Wang F (2025) Improving the properties of laccase through heterologous expression and protein engineering. Microorganisms 13(6):1422. https://doi.org/10.3390/microorganisms13061422

Guo N, Zheng J, Wu L, Tian J, Zhou H (2013) Engineered bifunctional enzymes of endo-1,4–xylanase/endo-1,4–mannanase were constructed for synergistically hydrolyzing hemicellulose. J Mol Catal B Enzym 97(2013):311–318. https://doi.org/10.1016/j.molcatb.2013.06.019

Hegazy UM, El-Khonezy MI, Shokeer A, Abdel-Ghany SS, Bassuny RI, Barakat AZ, Salama WH, Azouz RAM, Fahmy AS (2019) Revealing of a novel xylose-binding site of *Geobacillus stearothermophilus* xylanase by directed evolution. J Biochem 165:177–184. https://doi.org/10.1093/jb/mvy092

Henikoff S, Greene EA, Pietrokovski S, Bork P, Attwood TK, Hood L (1997) Gene families: the taxonomy of protein paralogs and chimeras. Science 278(5338):609–614. https://doi.org/10.1126/science.278.5338.609

Hollingsworth SA, Dror RO (2018) Molecular dynamics simulation for all. Neuron 99(6):1129–1143. https://doi.org/10.1016/j.neuron.2018.08.011

Hong S, Lee J, Cho K, Math R, Kim Y, Hong S, Cho Y, Cho S, Kim H, Yun H (2007) Construction of the bifunctional enzyme cellulase-β-glucosidase from the hyperthermophilic bacterium *Thermotoga maritima*. Biotechnol Lett 29:931–936. https://doi.org/10.1007/s10529-007-9334-5

Hossack E, Hardy F, Green A (2023) Building enzymes through design and evolution. ACS Catal 13(19):12436–12444. https://doi.org/10.1021/acscatal.3c02746

Hua YW, Chi MC, Lo HF, Hsu WH, Lin LL (2004) Fusion of *Bacillus stearothermophilus* leucine aminopeptidase II with the raw-starch-binding domain of *Bacillus* sp. strain TS-23 α-amylase generates a chimeric enzyme with enhanced thermostability and catalytic activity. J Ind Microbiol Biotechnol 31:273–277. https://doi.org/10.1007/s10295-004-0146-5

Illanes A, Cauerhff A, Wilson L, Castro GR (2012) Recent trends in biocatalysis engineering. Bioresour Technol 115:48–57

Jadaun JS, Narnoliya LK, Srivastava A, Singh SP (2020) Chimeric enzyme designing for the synthesis of multifunctional biocatalysts. In: Biomass, biofuels, biochemicals. Elsevier, pp 119–143. https://doi.org/10.1016/B978-0-12-819820-9.00008-9

Jumper J, Evans R, Pritzel A, Green T, Figurnov M, Ronneberger O, Tunyasuvunakool K, Bates R, Žídek A, Potapenko A, Bridgland A, Meyer C, Kohl SAA, Ballard AJ, Cowie A, Romera-Paredes B, Nikolov S, Jain R, Adler J, Back T, Petersen S, Reiman D, Clancy E, Zielinski M, Steinegger M, Pacholska M, Berghammer T, Bodenstein S, Silver D, Vinyals O, Senior AW, Kavukcuoglu K, Kohli P, Hassabis D (2021) Highly accurate protein structure prediction with AlphaFold. Nature 596(7873):583–589. https://doi.org/10.1038/s41586-021-03819-2

Jurich C, Shao Q, Ran X, Yang ZJ (2025) Physics-based modeling in the new era of enzyme engineering. Nat Comput Sci 5(4):279–291. https://doi.org/10.1038/s43588-025-00788-8

Karplus M, McCammon JA (2002) Molecular dynamics simulations of biomolecules. Nat Struct Biol 9(9):646–652. https://doi.org/10.1038/nsb0902-646

Khalili Ghadikolaei K, Akbari Noghabi K, Shahbani Zahiri H (2017) Development of a bifunctional xylanase-cellulase chimera with enhanced activity on rice and barley straws using a modular xylanase and an endoglucanase procured from camel rumen metagenome. Appl Microbiol Biotechnol 101(18):6929–6939. https://doi.org/10.1007/s00253-017-8430-2

Khamies M, Hagar M, Kassem TSE, Moustafa AHE (2024) Case study of chemical and enzymatic degumming processes in soybean oil production at an industrial plant. Sci Rep 14:4064. https://doi.org/10.1038/s41598-024-53865-9

Kim HM, Jung S, Lee KH, Song Y, Bae H-J (2015) Improving lignocellulose degradation using xylanase–cellulase fusion protein with a glycine–serine linker. Int J Biol Macromol 73:215–221. https://doi.org/10.1016/j.ijbiomac.2014.11.025

Koljak R, Boutaud O, Shieh BH, Samel N, Brash AR (1997) Identification of a naturally occurring peroxidase-lipoxygenase fusion protein. Science 277(5334):1994–1996. https://doi.org/10.1126/science.277.5334.1994

Kortemme T (2024) De novo protein design-From new structures to programmable functions. Cell 187(3):526–544. https://doi.org/10.1016/j.cell.2023.12.028

Kourist R, Jochens H, Bartsch S, Bornscheuer UT (2010) The role of active site residues in lipase substrate specificity. ChemBioChem 11(11):1635–1644

Krieger E, Joo K, Lee J, Lee J, Raman S, Thompson J, Tyka M, Baker D, Karplus K (2009) Improving physical realism, stereochemistry, and side-chain accuracy in homology modeling: Four approaches that performed well in CASP8. Proteins: Struct, Funct, Bioinform 77(S9):114–122. https://doi.org/10.1002/prot.22570

Kuhad RC, Singh A, Eriksson KE (1997) Microorganisms and enzymes involved in the degradation of plant fiber cell walls. Adv Biochem Eng Biotechnol 1997(57):45–125. https://doi.org/10.1007/BFb0102072. PMID: 9204751

Kuhlman B, Baker D (2000) Native protein sequences are close to optimal for their structures. Proc Natl Acad Sci 97(19):10383–10388

Kumar A (2020) Synthetic biology and future production of biofuels and high–value products. In: Kumar A, Yau YY, Ogita S, Scheibe R (eds) Climate change, photosynthesis and advanced biofuels. Springer, Singapore. https://doi.org/10.1007/978-981-15-5228-1_11

Kun RS, Garrigues S, Di Falco M, Tsang A, de Vries RP (2021) The chimeric GaaR- XlnR transcription factor induces pectinolytic activities in the presence of D-xylose in aspergillus Niger. Appl Microbiol Biotechnol 105:5553–5564. https://doi.org/10.1007/s00253-021-11428-2

Ladokhin AS, White SH (1999). Folding of amphipathic alpha-helices on membranes: energetics of helix formation by melittin. J Mol Biol 285(4):1363–9. https://doi.org/10.1006/jmbi.1998.2346. PMID: 9917380

Landwehr M, Carbone M, Otey CR, Li Y, Arnold FH (2007) Diversification of catalytic function in a synthetic family of chimeric cytochrome p450s. Chem Biol 14(3):269–78. https://doi.org/10.1016/j.chembiol.2007.01.009. PMID: 17379142; PMCID: PMC1991292

Leaver-Fay A, Tyka M, Lewis SM, Lange OF, Thompson J, Jacak R, Kaufman K, Renfrew PD, Smith CA, Sheffler W, Davis IW, Cooper S, Treuille A, Mandell DJ, Richter F, Ban YE, Fleishman SJ, Corn JE, Kim DE, Lyskov S, Berrondo M, Mentzer S, Popović Z, Havranek JJ, Karanicolas J, Das R, Meiler J, Kortemme T, Gray JJ, Kuhlman B, Baker D, Bradley P (2011) ROSETTA3: an object-oriented software suite for the simulation and design of macromolecules. Methods Enzymol 487:545–574. https://doi.org/10.1016/B978-0-12-381270-4.00019-6

Lee HL, Chang CK, Teng KH, Liang PH (2011) Construction and characterization of different fusion proteins XE "Fusion protein" between cellulases and beta-glucosidase to improve glucose production and thermostability. Bioresource Technol 102(4):3973–3976. https://doi.org/10.1016/j.biortech.2010.11.114

Leemhuis H, Kelly RM, Dijkhuizen L (2009) Engineering of cyclodextrin glucanotransferases and the impact for biotechnological applications. Appl Microbiol Biotechnol 85(4):823–835

Leman JK, Weitzner BD, Lewis SM, Adolf-Bryfogle J, Alam N, Alford RF, Aprahamian M, Baker D, Barlow KA, Barth P, Basanta B, Bender BJ, Blacklock K, Bonet J, Boyken SE, Bradley P, Bystroff C, Conway P, Cooper S, Correia BE, Coventry B, Das R, De Jong RM, DiMaio F, Dsilva L, Dunbrack R, Ford AS, Frenz B, Fu DY, Geniesse C, Goldschmidt L, Gowthaman R, Gray JJ, Gront D, Guffy S, Horowitz S, Huang P-S, Huber T, Jacobs TM, Jeliazkov JR, Johnson DK, Kappel K, Karanicolas J, Khakzad H, Khar KR, Khare SD, Khatib F, Khramushin A, King IC, Kleffner R, Koepnick B, Kortemme T, Kuenze G, Kuhlman B, Kuroda D, Labonte JW, Lai JK, Lapidoth G, Leaver-Fay A, Lindert S, Linsky T, London N, Lubin JH, Lyskov S, Maguire J, Malmstrom L, Marcos E, Marcu O, Marze NA, Meiler J, Moretti R, Mulligan VK, Nerli S, Norn C, ´O'Conchúir S, Ollikainen N, Ovchinnikov S, Pacella MS, Pan X, Park H, Pavlovicz RE, Pethe M, Pierce BG, Pilla KB, Raveh B, Renfrew PD, Burman SSR, Rubenstein A, Sauer MF, Scheck A, Schief W, Schueler-Furman O, Sedan Y, Sevy AM, Sgourakis NG, Shi L, Siegel JB, Silva D-A, Smith S, Song Y, Stein A, Szegedy M, Teets FD, Thyme SB, Wang RY-R, Watkins A, Zimmerman L, Bonneau R (2020) Macromolecular modeling and design in Rosetta: recent methods and frameworks. Nat Methods 17:665–680. https://doi.org/10.1038/s41592-020-0848-2

Liang J, Blumenthal RM (2013) Naturally-occurring, dually-functional fusions between restriction endonucleases and regulatory proteins. BMC Evol Biol 13(1):218

Lin Y, Jin W, Wang J, Cai Z, Wu S, Zhang G (2018) A novel method for simultaneous purification and immobilization of a xylanase-lichenase chimera via SpyTag/SpyCatcher spontaneous reaction. Enzym Microb Technol 115:29–36. https://doi.org/10.1016/j.enzmictec.2018.04.007

Liu Q, Xun G, Feng Y (2019) The state-of-the-art strategies of protein engineering for enzyme stabilization. Biotechnol Adv 37(4):530–537. https://doi.org/10.1016/j.biotechadv.2018.10.011. Epub 2018 Oct 26 PMID: 31138425

Liu M, Yang S, Long L, Cao Y, Ding S (2018a) Engineering a chimeric lipase-cutinase (Lip-Cut) for efficient enzymatic deinking of waste paper. BioRes 13(1):981–996

Liu M, Yang S, Long L, Wu S, Ding S (2017) The enzymatic deinking of waste papers by engineered bifunctional chimeric neutral lipase—endoglucanase. Bioresources 12. https://doi.org/10.15376/biores.12.3.6812-6831

Liu ZL, Li HN, Song HT et al (2018b) Construction of a trifunctional cellulase and expression in *Saccharomyces cerevisiae* using a fusion protein. BMC Biotechnol 18:43. https://doi.org/10.1186/s12896-018-0454-x

Long M (2000) A new function evolved from gene fusion. Genome Res 10(11):1655–1657

Long M, Langley CH (1993) Natural selection and the origin of jingwei, a chimeric processed functional gene in Drosophila. Science 260(5104):91–95. https://doi.org/10.1126/science.7682012

Luo M, Zhang M, Chi C, Chen G (2024) Affinity-assisted covalent self-assembly of PduQ-SpyTag and Nox-SpyCatcher to construct multi-enzyme complexes on the surface of magnetic microsphere modified with chelated Ni2+. Int J Biol Macromol 261:129365. https://doi.org/10.1016/j.ijbiomac.2024.129365

Ma Y, Zhang N, Vernet G, Kara S (2022) Design of fusion enzymes for biocatalytic applications in aqueous and non-aqueous media. Front Bioeng Biotechnol 10:944226. https://doi.org/10.3389/fbioe.2022.944226. PMID: 35935496; PMCID: PMC9354712

Mandai T, Fujiwara S, Imaoka S (2009) Construction and engineering of a thermostable self-sufficient cytochrome P450. Biochem Biophys Res Commun 384(1):61–65. https://doi.org/10.1016/j.bbrc.2009.04.064

Martins M, Dinamarco TM, Goldbeck R (2020) Recombinant chimeric enzymes for lignocellulosic biomass hydrolysis. Enzyme Microb Technol 140:109647. https://doi.org/10.1016/j.enzmictec.2020.109647

Méndez-Lucio O, Medina-Franco JL (2017) The many roles of molecular complexity in drug discovery. Drug Discov Today 22(1):120–126. https://doi.org/10.1016/j.drudis.2016.08.009

Minshull J, Stemmer WP (1999) Protein evolution by molecular breeding. Curr Opin Chem Biol 3(3):284–90. https://doi.org/10.1016/s1367-5931(99)80044-1

Mitkowski P, Jagielska E, Sabała I (2024) Engineering of chimeric enzymes with expanded tolerance to ionic strength. Microbiol Spectr 12:e03546-e3623. https://doi.org/10.1128/spectrum.03546-23

Monica P, Ranjan R, Kapoor M (2024) Lignocellulose-degrading chimeras: emerging perspectives for catalytic aspects, stability, and industrial applications. Renew Sust Ener Rev 199:114425. https://doi.org/10.1016/j.rser.2024.114425

Moretti R, Bender BJ, Allison B, Meiler J (2016) Rosetta and the design of ligand binding sites. Methods Mol Biol 47–62. https://doi.org/10.1007/978-1-4939-3569-7_4

Narnoliya LK, Jadaun JS, Singh SP (2018) Management of agro-industrial wastes with the aid of synthetic biology. In: Varjani S, Parameswaran B, Kumar S, Khare S (eds) Biosynthetic technology and environmental challenges. Energy, Environment, and Sustainability. Springer, Singapore. https://doi.org/10.1007/978-981-10-7434-9_2

Nath P, Dhillon A, Kumar K, Sharma K, Jamaldheen SB, Moholkar VS, Goyal A (2019) Development of bi-functional chimeric enzyme (CtGH1-L1-CtGH5-F194A) from endoglucanase (CtGH5) mutant F194A and β-1,4-glucosidase (CtGH1) from *Clostridium thermocellum* with enhanced activity and structural integrity. Bioresour Technol 282:494–501. https://doi.org/10.1016/j.biortech.2019.03.051

Nick Pace C, Martin Scholtz J, Grimsley GR (2014) Forces stabilizing proteins. FEBS Lett 588(14):2177–2184

Nixon AE, Ostermeier M, Benkovic SJ (1998) Hybrid enzymes: manipulating enzyme design. Trends Biotechnol 16(6):258–264. https://doi.org/10.1016/s0167-7799(98)01204-9

Ozer A, Uzuner U, Guler HI, Ay Sal F, Belduz AO, Deniz I, Canakci S (2018) Improved pulp bleaching potential of *Bacillus* subtilis WB800 through overexpression of three lignolytic enzymes from various bacteria. Biotechnol Appl Biochem 65(4):560–571. https://doi.org/10.1002/bab.1637

Parashar D, Satyanarayana T (2016) A chimeric α-amylase engineered from *Bacillus acidicola* and *Geobacillus thermoleovorans* with improved thermostability and catalytic efficiency. J Ind Microbiol Biotechnol 43:473–484. https://doi.org/10.1007/s10295-015-1721-7

Parashar D, Satyanarayana T (2017) Engineering a chimeric acid-stable α-amylase-glucoamylase (Amy-Glu) for one step starch saccharification. Int J Biol Macromolecules 99:274–281. https://doi.org/10.1016/j.ijbiomac.2017.02.083

Pathak P, Kaur P, Bhardwaj NK (2017) Chapter 6. Microbial enzymes for pulp and paper industry. In: Shukla P (ed) Microbial biotechnology: an interdiciplinary approach. CRC Press Taylor and Francis Group: Boca Raton, FL, USA, pp 163–240

Peng H, Li R, Li F, Zhai L, Zhang X, Xiao Y et al (2018) Extensive hydrolysis of raw rice starch by a chimeric α-amylase engineered with α-amylase (AmyP) and a starch-binding domain from

Cryptococcus Sp. S-2. Appl Microbiol Biotechnol 102:743–750. https://doi.org/10.1007/s00 253-017-8638-1

Pham LTM, Deng K, Choudhary H, Northen TR, Singer SW, Adams PD, Simmons BA, Sale KL (2024) An engineered Laccase from *Fomitiporia mediterranea* accelerates lignocellulose degradation. Biomolecules 14(3):324. https://doi.org/10.3390/biom14030324

Piatkowska EM, Naseeb S, Knight D, Delneri D (2013) Chimeric protein complexes in hybrid species generate novel phenotypes. PLoS Genet 9(10):e1003836. https://doi.org/10.1371/jou rnal.pgen.1003836

Pinheiro MP, Reis RAG, Dupree P, Ward RJ (2021) Plant cell wall architecture guided design of CBM3-GH11 chimeras with enhanced xylanase activity using a tandem repeat left-handed β-3-prism scaffold. Comput Struct Biotechnol J 19:1108–1118. https://doi.org/10.1016/j.csbj.2021. 01.011

Planas-Iglesias J, Marques SM, Pinto GP, Musil M, Stourac J, Damborsky J, Bednar D (2021a) Computational design of enzymes for biocatalysis. Biotechnol Adv 49:107733. https://doi.org/ 10.1016/j.biotechadv.2021.107733

Planas-Iglesias J, Marques SM, Pinto GP, Musil M, Stourac J, Damborsky J, Bednar D (2021b) Computational design of enzymes for biotechnological applications. Biotechnol Adv 47:107696. https://doi.org/10.1016/j.biotechadv.2021.107696

Plummer SM, Plummer MA, Merkel PA, Hagen M, Biddle JF, Waidner LA (2016) Using directed evolution to improve hydrogen production in chimeric hydrogenases from *Clostridia* species. Enzyme Microb Technol 93–94:132–141. https://doi.org/10.1016/j.enzmictec.2016.07.011

Press RD, Kamel-Reid S, Ang D (2013) BCR-ABL1 RT-qPCR for monitoring the molecular response to tyrosine kinase inhibitors in chronic myeloid leukemia. J Mol Diagn 15(5):565–576. https://doi.org/10.1016/j.jmoldx.2013.04.007

Pulp and Paper Enzymes Market, Global Outlook and Forecast 2025–2032 https://chemicalr esearchinsight.com/2025/10/17/pulp-and-paper-enzymes-market-global-outlook-and-forecast-2025-2

Ravalason H, Herpoël-Gimbert I, Record E, Bertaud F, Grisel S, de Weert S, van den Hondel CA, Asther M, Petit-Conil M, Sigoillot JC (2009) Fusion of a family 1 carbohydrate binding module of *Aspergillus niger* to the *Pycnoporus cinnabarinus* laccase for efficient softwood kraft pulp biobleaching. J Biotechnol 142(3–4):220–6. https://doi.org/10.1016/j.jbiotec.2009.04.013

Ribeiro LF, Furtado GP, Lourenzoni MR, Costa-Filho AJ, Santos CR, Nogueira SC, Betini JA, Polizeli Mde L, Murakami MT, Ward RJ (2011) Engineering bifunctional laccase-xylanase chimeras for improved catalytic performance. J Biol Chem 286(50):43026–38. https://doi.org/ 10.1074/jbc.M111.253419

Ribeiro LF, Tullman J, Nicholes N, Silva SR, Vieira DS, Ostermeier M, Ward RJ (2016) A xylose-stimulated xylanase–xylose binding protein chimera created by random nonhomologous recombination. Biotechnol Biofuels 9:119

Rizk M, Elleuche S, Antranikian G (2015) Generating bifunctional fusion enzymes composed of heat-active endoglucanase (Cel5A) and endoxylanase (XylT). Biotechnol Lett 37:139–145. https://doi.org/10.1007/s10529-014-1654-7

Robinson CR, Sauer RT (1998) Optimizing the stability of single-domain proteins through linker design. Proc Natl Acad Sci USA 95(11):5929–5934. https://doi.org/10.1073/pnas.95.11.5929

Rogers RL, Hartl DL (2012) Chimeric genes as a source of rapid evolution in *Drosophila melanogaster*. Mol Biol Evol 29(2):517–529. https://doi.org/10.1093/molbev/msr184

Romero PA, Arnold FH (2009) Exploring protein fitness landscapes by directed evolution. Nat Rev Mol Cell Biol 10(12):866–876. https://doi.org/10.1038/nrm2805

Roy S (1999) Multifunctional enzymes and evolution of biosynthetic pathways: retro-evolution by jumps. Proteins Struct Funct Bioinform 37(2):303–309

Saadat F (2017) A review on chimeric xylanases: methods and conditions. 3 Biotech 7(1):67

Sakhuja D, Ghai H, Rathour RK, Kumar P, Bhatt AK, Bhatia RK (2021) Cost-effective production of biocatalysts using inexpensive plant biomass: a review. 3 Biotech 11(6):280

Sali A, Blundell TL (1993) Comparative protein modelling by satisfaction of spatial restraints. J Mol Biol 234(3):779–815

Shahid S, Batool S, Khaliq A, Ahmad S, Batool H, Sajjad M, Akhtar MW (2023) Improved catalytic efficiency of chimeric xylanase 10B from *Thermotoga petrophila* RKU1 and its synergy with cellulases. Enzym Microb Technol 166:110213. https://doi.org/10.1016/j.enzmictec.2023.110213

Shapiro JA (2002) Genome organization and reorganization in evolution. Ann N Y Acad Sci 981(1):111–134

Singh R, Kumar M, Mittal A, Mehta PK (2016) Microbial enzymes: Industrial progress in 21st century. 3 Biotech 6(2):174. https://doi.org/10.1007/s13205-016-0485-8

Son A, Park J, Kim W, Yoon Y, Lee S, Park Y, Kim H (2024) Revolutionizing molecular design for innovative therapeutic applications through artificial intelligence. Molecules 29(19):4626. https://doi.org/10.3390/molecules29194626

Song L, Dumon C, Siguier B, Andre I, Eneyskaya E, Kulminskaya A et al (2014) Impact of an N-terminal extension on the stability and activity of the GH11 xylanase from *Thermobacillus xylanilyticus*. J Biotechnol 174(1):64–72

Sørensen HP, Mortensen KK (2005) Advanced genetic strategies for recombinant protein expression in *Escherichia coli*. J Biotechnol 115(2):113–128. https://doi.org/10.1016/j.jbiotec.2004.08.004

Sorrentino I, Giardina P, Piscitelli A (2019) Development of a biosensing platform based on a laccase-hydrophobin chimera. Appl Microbiol Biotechnol 103(7):3061–3071

Strohl W (2017) Chimeric genes, proteins. In: Reference module in life sciences. Elsevier. https://doi.org/10.1016/B978-0-12-809633-8.20482-3

Sumbalova L, Stourac J, Martinek T, Bednar D, Damborsky J (2018) HotSpot wizard 3.0: web server for automated design of mutations and smart libraries based on sequence input information. Nucleic Acids Res 46:W356–W362. https://doi.org/10.1093/nar/gky417

Sun JY, Liu MQ, Xu YL, Xu ZR, Pan L, Gao H (2005) Improvement of the thermostability and catalytic activity of a mesophilic family 11 xylanase by N-terminus replacement. Protein Expr Purif 42(1):122–130

Sun Z, Liu Q, Qu G, Feng Y, Reetz MT (2019) Utility of B-factors in protein science: interpreting rigidity, flexibility, and internal motion and engineering thermostability. Chem Rev 119(3):1626–1665. https://doi.org/10.1021/acs.chemrev.8b00290

Suurnakki A, Tenkanen M, Buchert J, Viikari L (1997) Hemicellulases in the bleaching of chemical pulps. Adv Biochem Eng/Biotechnol 57:261–287

Thulasiram HV, Erickson HK, Poulter CD (2007) Chimeras of two isoprenoid synthases catalyze all four coupling reactions in isoprenoid biosynthesis. Science 316(5821):73–76

Tokuriki N, Tawfik DS (2009) Protein dynamism and evolvability. Science 324(5924):203–207

Torng W, Altman RB (2019) High precision protein functional site detection using 3D convolutional neural networks. Bioinformatics 35(9):1503–1512. https://doi.org/10.1093/bioinformatics/bty813

Turchetto-Zolet AC, Margis-Pinheiro M, Margis R (2009) The evolution of pyrroline-5-carboxylate synthase in plants: a key enzyme in proline synthesis. Mol Genet Genomics 281(1):87–97. https://doi.org/10.1007/s00438-008-0396-4

Turner NJ (2009) Directed evolution drives the next generation of biocatalysts. Nat Chem Biol 5(8):567–573

van Rosmalen M, Krom M, Merkx M (2017) Tuning the flexibility of glycine–serine linkers to allow rational design of multidomain proteins. Biochemistry 56(50):6565–6574. https://doi.org/10.1021/acs.biochem.7b00902

Varadi M, Anyango S, Deshpande M, Nair S, Natassia C, Yordanova G, Velankar S (2022) AlphaFold Protein Structure Database: Massively expanding the structural coverage of protein-sequence space with high-accuracy models. Nucleic Acids Res 50(D1):D439–D444. https://doi.org/10.1093/nar/gkab1061

Viikari L, Ranua M, Kantelinen A, Sundquist J, Linko M (1986) Bleaching with enzymes. In: Proceedings of 3rd international conference on biotechnology in the pulp and paper industry, 67–9. STFI, Stockholm

Walsh G (2010) Biopharmaceutical benchmarks. Nat Biotechnol 28(9):917–924

Walsh G (2018) Biopharmaceutical benchmarks 2018. Nat Biotechnol 36(12):1136–1145. https://doi.org/10.1038/nbt.4305

Wang H, Qi X, Gao S, Zhang Y, An Y (2022) Biochemical characterization of an engineered bifunctional xylanase/feruloyl esterase and its synergistic effects with cellulase on lignocellulose hydrolysis. Bioresour Technol 355:127244. https://doi.org/10.1016/j.biortech.2022.127244.

Wang R, Yang J, Jang JM, Liu J, Zhang Y, Liu L, Yuan H (2020) Efficient ferulic acid and xylo-oligosaccharides production by a novel multi-modular bifunctional xylanase/feruloyl esterase using agricultural residues as substrates. Bioresour Technol 297:122487. https://doi.org/10.1016/j.biortech.2019.122487

Wang X, Jiang Y, Liu H et al (2023) Research progress of multi-enzyme complexes based on the design of scaffold protein. Bioresour Bioprocess 10:72. https://doi.org/10.1186/s40643-023-00695-8

Wilson DB, Kostylev M (2012) Cellulase processivity. Methods Mol Biol 908:93–99

Wriggers W, Schulten K (1999) Investigating a back door mechanism of actin phosphate release by steered molecular dynamics. Proteins: Struct, Funct, Bioinform 35(3):262–273. https://doi.org/10.1002/(SICI)1097-0134(19990601)35:3<262::AID-PROT6>3.0.CO;2-R

Wu B, Yu Q, Chang S, Pedroso MM, Gao Z, He B, Schenk G (2019) Expansin assisted bio-affinity immobilization of endoxylanase from *Bacillus subtilis* onto corncob residue: characterization and efficient production of xylooligosaccharides. Food Chem 282:101–108. https://doi.org/10.1016/j.foodchem.2019.01.004

Yadav K, Yadav S, Tanveer A, Gupta S, Dwivedi S, Yadav D (2023) Innovations in papermaking using enzymatic intervention: an ecofriendly approach. Cellulose 30(15):7393–7425. https://doi.org/10.1007/s10570-023-05333-2

Yakubu A, Saikia U, Vyas A (2019) Microbial enzymes and their application in pulp and paper industry. In: Yadav A, Singh S, Mishra S, Gupta A (eds) Recent advancement in white biotechnology through fungi. Fungal Biology. Springer, Cham. https://doi.org/10.1007/978-3-030-25506-0_12

Yang J, Lal RG, Bowden JC, Astudillo R, Hameedi MA, Kaur S, Hill M, Yue Y, Arnold FH (2025) Active learning-assisted directed evolution. Nat Commun 16(1):714. https://doi.org/10.1038/s41467-025-55987-8

Yang HQ, Liu L, Xu F (2016) The promises and challenges of fusion constructs in protein biochemistry and enzymology. Appl Microbiol Biotechnol 100(19):8273–8281. https://doi.org/10.1007/s00253-016-7795-y

Yang J, Yan R, Roy A, Xu D, Poisson J, Zhang Y (2015) The I-TASSER Suite: protein structure and function prediction. Nat Methods 12(1):7–8. https://doi.org/10.1038/nmeth.3213

Yang KK, Wu Z, Arnold FH (2022) Machine-learning-guided directed evolution for protein engineering. Nat Methods 16(7):687–694

Yang M, Li J, Wang S, Zhao F, Zhang C, Zhang C, Han S (2023) Status and trends of enzyme cocktails for efficient and ecological production in the pulp and paper industry. J Clean Prod 418:138196

Yang Y, Zhang C, Lu H, Wu Q, Wu Y, Li W, Li X (2024) Improvement of thermostability and catalytic efficiency of xylanase from Myceliophthora thermophilar by N-terminal and C-terminal truncation. Front Microbiol 15. https://doi.org/10.3389/fmicb.2024.1385329

Yin X, Li JF, Wang JQ, Tang CD, Wu MC (2013) Enhanced thermostability of a mesophilic xylanase by N-terminal replacement designed by molecular dynamics simulation. J Sci Food Agric 93(12):3016–3023. https://doi.org/10.1002/jsfa.6134

You C, Percival Zhang YH (2014) Constructing biofuel-producing microorganisms by synthetic biology and metabolic engineering. Biotechnol Adv 32(5):817–831

Yu K, Liu C, Kim BG, Lee DY (2015) Synthetic fusion protein design and applications. Biotechnol Adv 33(1):155–164. https://doi.org/10.1016/j.biotechadv.2014.11.005

Zhang J, Dean AM, Brunet F, Long M (2004) Evolving protein functional diversity in new genes of *Drosophila*. Proc Natl Acad Sci U S A 101(46):16246–16250. https://doi.org/10.1073/pnas.0407066101

Zhang J, Moilanen U, Tang M, Viikari L (2013) The carbohydrate-binding module of xylanase from *Nonomuraea flexuosa* decreases its non-productive adsorption on lignin. Biotechnol Biofuels 6:18. https://doi.org/10.1186/1754-6834-6-18

Zhang P, Bao Y, Higgins L, Xu SY (2007) Rational design of a chimeric endonuclease targeted to NotI recognition site. Protein Eng des Sel 20(10):497–504. https://doi.org/10.1093/protein/gzm049

Zhao H, Ding W, Zang J, Yang Y, Liu C, Hu L, Chen Y, Liu G, Fang Y, Yuan Y, Lin S (2021) Directed-evolution of translation system for efficient unnatural amino acids incorporation and generalizable synthetic auxotroph construction. Nat Commun 12:7039. https://doi.org/10.1038/s41467-021-27399-x

Further Reading

Amitai G, Shemesh A, Sitbon E, Shklar M, Netanely D, Venger I, Pietrokovski S (2004) Network analysis of protein structures identifies functional residues. J Mol Biol 344(4):1135–1146. https://doi.org/10.1016/j.jmb.2004.09.072

Angural S, Jassal S, Warmoota R, Rana M, Puri N, Gupta N (2023) An integrated approach for pulp biobleaching: application of cocktail of enzymes. Environ Sci Pollut Res 30:57155–57163

Balda S, Sharma A, Gupta N, Capalash N, Sharma P (2022) Deinking of old newsprint (ONP) pulp with an engineered laccase: a greener approach for paper recycling. Biomass Convers Biorefinery 14:3965–3974

Blow DM (2000) So do we understand how enzymes work? Structure 8(2):R77–R81

Demuner BJ, Pereira Junior N, Antunes A (2011) Technology prospecting on enzymes for the pulp and paper industry. J Technol Manag Innov 6:148–158

Dixit M, Gupta GK, Pathak P, Bhardwaj NK, Shukla P (2022) An efficient endoglucanase and lipase enzyme consortium (ELEC) for deinking of old newspaper and ultrastructural analysis of deinked pulp. Biomass Conversion and Biorefinery 15:22003–22011

Fowler DM, Araya CL, Fleishman SJ, Kellogg EH, Stephany JJ, Baker D, Fields S (2010) High-resolution mapping of protein sequence–function relationships. Nat Methods 7(9):741–746. https://doi.org/10.1038/nmeth.1492

Höcker B (2013) Engineering chimeric proteins from fold fragments: "Hopeful monsters" in protein design. Biochem Soc Trans 41(5):1137–1140

https://infinitabiotech.com/blog/applications-of-enzymes-in-paper-and-pulp-industries/

https://www.creative-enzymes.com/resource/application-of-enzymes-in-pulp-and-paper-industry_64.html

Khersonsky O, Lipsh R, Avizemer Z, Ashani Y, Goldsmith M, Leader H, Fleishman SJ (2018) Automated design of efficient and functionally diverse enzyme repertoires. Mol Cell 72(1):178-186.e5. https://doi.org/10.1016/j.molcel.2018.08.033

Kries H, Niquille DL, Hilvert D (2015) A subdomain swap strategy for reengineering nonribosomal peptides. Chem Biol 22(5):640–648—Experimental subdomain-swap strategy for NRPS adenylation domains (good direct example of domain/subdomain swapping in biosynthetic enzymes)

Kumar V, Marín-Navarro J, Shukla P (2016) Thermostable microbial xylanases for pulp and paper industries: trends, applications and further perspectives. World J Microbiol Biotechnol 32(2):34. https://doi.org/10.1007/s11274-015-2005-0

Levasseur A, Navarro D, Punt PJ, Belaïch JP, Asther M, Record E (2005) Construction of engineered bifunctional enzymes and their overproduction in *Aspergillus niger* for improved enzymatic tools to degrade agricultural by-products. Appl Environ Microbiol 71(12):8132–40. https://doi.org/10.1128/AEM.71.12.8132-8140.2005

Martins M, Dinamarco TM, Goldbeck R (2020) Recombinant chimeric enzymes for lignocellulosic biomass hydrolysis. Enzyme Microb Technol 140:109647

Moraïs S, Barak Y, Caspi J, Hadar Y, Lamed R, Shoham Y, Wilson DB, Bayer EA (2010) Cellulase-xylanase synergy in designer cellulosomes for enhanced degradation of a complex cellulosic substrate. mBio. 1(5):e00285–10. https://doi.org/10.1128/mBio.00285-10

Nagano N, Orengo CA, Thornton JM (2002) One fold with many functions: the evolutionary relationships between TIM barrel families based on their sequences, structures, and functions. J Mol Biol 321(5):741–765. https://doi.org/10.1016/S0022-2836(02)00649-6

Tizei PA, Csibra E, Torres L, Pinheiro VB (2016) Selection platforms for directed evolution in synthetic biology. Biochem Soc Trans 44(4):1165–1175. https://doi.org/10.1042/BST20160061

Wei S, Liu K, Ji X, Wang T, Wang R (2021) Application of enzyme technology in biopulping and biobleaching. Cellulose 28:10099–10116

Index

A

Accessory domains, 84
Accessory esterase, 84
Acetyl xylan esterase, 29
Active site, 40, 51, 61, 65, 69, 71–75,
    77–79, 85, 101, 102
Adhesives, 5, 13, 25, 27, 36
Adsorbable Organic Halides (AOX), 15, 87
Agricultural residues, 5, 6
Agricultural waste valorization, 93, 103
AI-assisted workflows, 83
AI-based predictive models, 71
AI-driven computational design, 40
Alcohol dehydrogenase, 23, 70
Alcohol oxidase, 23, 25
Aldehyde dehydrogenase, 70
Alkali-resistant, 27, 28
α-Amylases, 22, 25, 66
α/β-hydrolase folds, 75
α-D-glucuronidase, 22, 24
α-glucuronidase, 22, 24
α-L-arabinofuranosidase, 24
Amylase, 21, 22, 25, 28, 36–40, 66
Animal nutrition, 4
Arabinofuranosidase, 29, 59, 66, 92, 96
Artificial Intelligence (AI), 16, 41, 79, 83,
    107
Aryl-alcohol dehydrogenases, 23
Aryl-alcohol oxidase, 23
Ash, 15
Aspergillus niveus, 60, 63
Aspergillus sp, 29
Autodock, 79
Auxiliary technologies, 40

B

Bacillus amyloliquefaciens, 31
Bacillus firmus, 30
Bacillus halodurans, 29
Bacillus nealsonii, 33
Bacillus pumilus, 31, 59
Bacillus sp, 29, 33
Bacillus subtilis, 62, 63, 71, 89, 96, 97,
    99–101
Bacillus tequilensis, 32
Bagasse, 14, 39, 66, 91, 95
Bamboo, 14, 27, 34, 39, 40
Bamboo kraft pulp, 34, 40
Bark, 15
Beating, 22–24, 27
Best Available Techniques (BAT), 15
β-D-xylosidase, 22, 24
β-1,4-endoglucanase, 62, 63
β-1,3-1,4-glucanases, 63
β-1,4-glucosidase, 60, 94
β-glucosidase, 22, 24, 27, 28, 60, 62, 63,
    71, 81, 84, 94, 95
β-glucosidase–xylanase, 63
β-xylanase, 62, 63, 96
β-xylosidase, 62, 63, 90, 92, 96
β-xylosidase–endoxylanase chimeras, 63
Bifunctional chimeras, 63
Biobleaching, 1, 19, 22, 23, 27–29, 34, 39,
    88, 100, 101, 106
Biodeinking, 22, 26, 28, 39
Bioenergy enzymes, 84
Bioethanol, 17
Biofuels, 4, 6, 7, 53, 65, 80, 90, 92, 93, 102,
    103
Bioinformatics, 51, 54, 68, 70
Biological Oxygen Demand (BOD), 15

P. Bajpai, *Chimeric Enzymes Showing Promise in Pulp and Paper Sector*,
SpringerBriefs in Molecular Science, https://doi.org/10.1007/978-3-032-18003-2

Biological treatments, 19
Biomass conversion, 1, 7, 70, 94, 103
Bioplastics, 6, 17
Biopulping, 19, 22, 23, 26–28, 37, 90, 100
Biorefinery, 6, 7, 11, 17, 71, 100
Biorefinery integration, 11, 17
Bioremediation, 4, 81
Biosensor, 66, 84
Birch, 14
Birchwood xylan, 63
Blastp, 62
Bleachability, 23
Bleaching, 1, 3–7, 11, 15, 20, 22–26, 36,
    38–40, 49, 59, 87, 89, 94, 100, 103,
    105, 106
Blue Angel, 16

C
Carbohydrate Binding Module (CBM), 25,
    26, 62, 63, 81, 82, 84, 88, 92, 93,
    100, 101, 103
Carbon Capture and Storage (CCS), 17
Carbon neutrality, 2, 17
Cardanol, 28, 40
Cardboard, 14
Cascade chimeras, 85
Catalytic domain, 51, 60, 62, 63, 65–69, 73,
    74, 77, 82, 84, 88, 91, 94, 105
Cationic polyacrylamide, 28
Cationic polymer, 24, 34
Cationic starch, 35
CBM44-XEGA, 60, 63
CBM-Enhanced Chimeric Constructs, 63
CcBglA β-glucosidase, 63
Cellobiose, 23, 24
Cellobiose dehydrogenase, 23
Cellooligosaccharides, 24
Cellulase, 3, 5, 7, 15, 21, 22, 24–28, 36,
    38–40, 53, 54, 63, 70, 71, 74, 80–82,
    84, 87, 88, 90–92, 95, 102–106
Cellulase chimeras, 80, 82
Cellulase domains, 105
Cellulase–lpmo fusions, 103
Cellulase-xylanase, 5, 7, 71
Cellulolytic systems, 84
Cellulose, 6, 13, 14, 22–26, 28, 37, 53, 60,
    81, 87, 90–92, 94, 102, 105
Cellulose-binding modules, 92
Cellulose nanofibers, 13
Cellulosome, 65, 82, 84, 90
Chemical additives, 21, 34–36, 40
Chemical dispersants, 5

Chemical Oxygen Demand (COD), 15
Chemical pretreatment, 40, 101
Chemical pulping, 14
Chemical treatment, 4, 5, 14, 23, 24, 87
Chimera protein, 51, 58
Chimeric, 1, 3–8, 25, 49–51, 53–55, 58–82,
    84–87, 89–101, 103–108
Chimeric antibodies, 53, 64
Chimeric cellulases, 74, 84, 92
Chimeric cellulase-xylanase, 5
Chimeric constructs, 4, 8, 63, 75, 80, 92,
    94, 100, 102, 103
Chimeric cytokines, 65
Chimeric enzyme engineering, 72, 74, 75,
    77, 78, 85, 92
Chimeric enzymes, 1, 3–8, 49, 51, 53, 54,
    57, 59–63, 65–81, 84, 85, 87–95,
    100, 101, 103–108
Chimeric gene, 50, 71
Chimeric hydrogenases, 85
Chimeric lipase-esterase enzyme, 5
Chimeric monoclonal antibodies, 64
Chimeric oxidoreductases, 71
Chimeric proteins, 50, 51, 53, 54, 64–66, 91
Chimeric xylanases, 4, 59
Chimerization, 54, 59, 61
Chinese hamster ovary, 51, 54
Chipping, 14
Chlorinated organics, 4, 87
Chlorine-based bleaching agents, 5, 100
Chlorine dioxide, 15, 34, 87, 105
Chromogenic groups, 22
Circular dichroism spectroscopy, 71, 73
Circular economy, 5, 6, 13, 14, 17, 41, 106
Climate change, 2
Clostridium thermocellum CBM44, 63
Clustalw, 62
Cluster Rule, 15
Compostability, 17
Composting, 15
Computational design, 40, 71, 73, 81, 83,
    107, 108
Computational modeling, 68, 70, 72,
    77–80, 82, 102
Computational protein modeling, 102
Computational simulation, 73, 74, 79
Conformational coupling, 72
Contaminant removal, 39, 88
Contaminants, 5, 36, 37, 39, 88
Conventional deinking, 5, 91
Cooking liquor, 14
Corn cob, 63
Counter-current flow systems, 15

COVID-19 vaccines, 65
Cryo-electron microscopy, 73
C-terminal domain, 59
C-terminal linker, 100
C. thermocellum, 60, 63, 94, 96, 97
Customizability, 7
Cutinases, 25
Cyanide dehydratase, 56, 59

D
DBTL cycle, 85
Debarking, 14, 37
Degumming, 91
Deinking, 1, 3, 5, 7, 19–25, 28, 34, 36–40,
    49, 53, 87, 89, 91, 103, 105, 106, 108
Depolymerization, 6, 23, 65, 92
Designer enzymes, 51
Digitalization, 12, 13, 16, 107
Directed evolution, 4, 40, 51, 60, 68, 71,
    73, 74, 77, 81, 83, 85–87, 102, 107
Dissolving pulp, 3, 15, 20, 25, 28, 35, 37, 39
Disulfide bonds, 59, 61, 82
Diverse feedstocks, 5
Dockerin-fused laccase, 90
Domain insertion, 51–53, 66, 91, 94, 98
Domain selection, 62, 67, 81, 82
Domain shuffling, 68
Domain swapping, 40, 69–72, 78
Dye decolorization, 84

E
Ecolabels, 16
Effluent toxicity, 4, 5, 24, 90, 103, 108
Elemental chlorine-free bleaching, 15
Elemental Chlorine-Free (ECF), 15, 39
Endo-(1,4)-β-D-glucanase, 24
Endo-1,4-β D-xylanase, 22
Endo-1,4-β-xylanase-β-xylosidase, 96
Endo5A, 62, 63, 98
Endogalacturonase, 92
Endoglucanase, 22, 25, 60–63, 84, 90, 91,
    94, 98
Endoglucanase (Endo5A)-Xyl11D, 98
Endoxylanase/endomannase chimera, 66
Endoxylanase–β-xylosidase chimera, 63
Endo-xylanases, 24, 62, 63, 66, 92, 97, 99
Energy efficiency, 16, 41, 91, 106, 108
Energy recovery, 15
Energy saving, 37, 106
Engineered laccases, 84
Engineered proteins, 3, 88, 101, 102
Enterobacter ludwigii, 30

Environmental footprint, 4, 5
Environmental impact, 7, 11, 15, 19, 90,
    100, 106, 107
Environmental pollution, 4, 105
Environmental Protection Agency (EPA),
    15, 102
Enzymatic deconstruction, 90
Enzymatic prebleaching, 100
Enzymatic pulping, 14, 26
Enzymatic refining/beating, 27
Enzymatic treatments, 4, 15, 19, 88, 105
Enzyme-assisted pulping, 7, 13
Enzyme-based catalysis, 3
Enzyme-based innovation, 3
Enzyme cocktails, 4, 19, 21, 26, 27, 35, 36,
    41, 65, 84, 102, 103
Enzyme engineering, 6, 7, 19, 40, 78, 83,
    102, 103, 107
Enzymes, 1, 3–8, 14, 15, 19–23, 25–29, 31,
    34, 36–41, 49–51, 53–55, 57–81, 83,
    84, 86–95, 98, 100–108
Enzyme treatments, 3, 4, 27, 28, 100
Escherichia coli, 51, 54, 71, 76, 80, 89, 92,
    97, 101
Espript, 62
Esterases, 5, 7, 19, 22, 24, 25, 29, 35–38,
    63, 74, 84, 93, 95, 97, 98, 103
Eucalyptus, 14, 25, 29, 39
Eucalyptus kraft pulp, 25, 29
Eutectic solvents, 54, 59
Exo-(1,4)-β-D-glucanase, 24
Exoglucanases, 22, 84, 90
Expansins, 25
Expression host, 62, 81, 82, 84, 95–99
Expression vector, 51, 54
Extremophile enzymes, 64

F
Fermentation technologies, 6, 107
Ferredoxin-like folds, 75
Feruloyl esterase, 25, 35, 63, 84, 95, 97
Feruloyl esterase–endoxylanase chimera,
    63
Fiber disintegration, 27
Fiber flexibility, 35, 40, 88
Fiber modification, 1, 4, 20, 21, 37–39, 88,
    90, 103, 105, 106
Fibrillation, 24, 27, 105, 106, 108
Fibrolytic enzymes, 3
Flexible linker, 80, 100
FoldX, 61, 77
Food packaging, 15

Food processing, 4, 92
Forest Stewardship Council (FSC), 16
Functional coatings, 15
Fungal oxidases, 25
Fusing domains, 5, 52, 82, 86
Fusion protein, 50, 51, 53, 54, 58, 64, 86, 90

G
Gaar–xlnr chimera, 93
Gene duplication, 50, 68
Generally Recognized As Safe (GRAS), 89
Genes, 50–54, 58, 65, 68, 92, 93
Genetically modified enzymes, 65
Genomes, 58, 62
GH10 xylanase, 59
Ginzu technique, 62
Glucoamylase, 55
Glucose oxidase, 23, 74
Glycine-rich linkers, 70, 82
Glycoproteins, 23
Glycosylation, 62, 81, 82, 84, 86, 89
Glycosyl hydrolases, 70
Glyoxal oxidase, 23
Gram-positive bacterium, 89
Greenhouse gas emissions, 2, 6, 106
GROMACS, 80

H
Hardwood kraft pulp, 34
Hardwoods, 5, 6, 14, 27, 34
Hemicellulase, 7, 15, 21, 24, 27, 36, 38, 84,
90, 106, 108
Hemicellulose, 4–7, 17, 23, 24, 27, 39, 66,
84, 90, 92, 105
Hemoperoxidase, 54
High-throughput screening, 71, 73, 78, 83,
86
Homology modeling, 70, 72, 73, 76, 79
Hotspot Wizard, 61
Human embryonic kidney-293, 54
Hybrid enzymes, 4, 23, 60, 66
Hydrogenase, 84, 85, 87
Hydrogen peroxide, 15, 34, 35
Hydrolase Engineering, 74
Hydrolases, 21, 27, 70, 71, 84
Hydrolytic enzymes, 25, 90
Hygiene and tissue products, 17

I
Immobilization engineering, 4

Ink, 5, 22, 25, 26, 28, 34, 37, 87, 88, 98,
106, 108
Insect cell, 51, 54, 77
Insertional fusion, 92
Integrated Pollution Prevention and Control
(IPPC) Directive, 15
Interdomain epistasis, 86
Ionic surfactant, 54, 59
I-TASSER, 78

K
Kenaf, 14
Kraft process, 14
Kraft pulp, 15, 25, 27, 29, 34, 40, 89, 101
Kraft softwood pulp, 30

L
Laccase, 19, 23, 24, 27–29, 32, 34, 36–40,
55, 59, 63, 66, 67, 84, 87, 88, 90–94,
100, 101, 103, 105
Laccase-CBM, 101
Laccase/endoxylanase, 66
Laccase–β-1,3-1,4-glucanase chimera, 63
Laccase–endoxylanase chimera, 92
Laccase/glucanase chimera, 66, 91
Laccase Mediator System (LMS), 23, 34,
36, 39
Laccase–xylanase chimeras, 94
Laccase-xylanase treatment, 27
Lecitase ultra, 91
Lichenase–xylanase chimera, 63
Lignin, 3, 4, 6, 13–15, 17, 19, 21–28, 37,
39, 84, 87, 88, 90, 91, 94, 100, 105,
108
Ligninases, 25, 27
Lignin auxiliary enzymes, 23
Lignin-based adhesives, 13
Lignin degradation, 21–24, 88, 100
Lignin degrading enzymes, 23
Lignin modification, 39, 84, 94
Lignin peroxidase, 21, 23, 27, 100
Lignin removal, 4, 26, 105, 108
Lignin-targeting chimeras, 84
Lignocellulosic biomass, 1, 4, 6, 14, 53, 54,
66, 70, 81, 84, 90, 93, 94, 103
Lignocellulosic biomass deconstruction,
81, 84, 94
Linker, 51, 54, 58, 62, 63, 70, 72–75,
78–87, 95–100, 103
Linker design, 73, 79, 82, 84
Linker engineering, 73, 80, 81, 83
Lipase-cutinase, 25, 53, 96

Lipase-Endoglucanase, 91, 98
Lipase-endoglucanase chimera, 91
Lipase-esterase, 5, 7
Lipase–esterase chimeras, 103
Lipase–phospholipase a1 chimera, 91
Lipases, 5, 22, 25, 34, 36–39, 53, 74, 88, 91, 93, 96, 103, 106, 108
Lip-Cut, 25, 53
Lipolytic enzymes, 21, 22, 25, 36
Lipoxygenase, 50, 54
Loosenin, 26
Lytic polysaccharide monooxygenase, 21, 22, 25

M
Machine learning, 16, 40, 41, 70, 74, 79, 83, 87, 107, 108
Machine-learning models, 77
Machine runnability, 5
Mammalian cell lines: HEK293, 54
Mammalian cells, 51, 54, 62, 81, 82
Manganese peroxidases, 21, 23
Mannanase, 19, 27, 30, 32, 33, 35, 37, 61, 63, 92, 95–97, 100
Mannanase-xylanase, 95, 96
Mannanase–xylanase chimera, 63
Mechanical pulping, 14, 103
Mechanical refining, 14, 40, 105, 106
Mediators, 23, 34, 36, 37
Mesophilic, 59, 60, 69, 75, 77
Mesophilic binding module, 77
Mesophilic xylanases, 59
Metabolomics, 107
Methylotrophic yeast, 89
Michaelis-Menten constant, 101
Microbial pulping, 26
Mixed office waste, 34
Mixed waste paper, 34
Mixed wood, 32, 33, 35
Mixed wood pulp, 39
MM/PBSA, 79
Modular domain engineering, 92
Molecular docking, 70, 73
Molecular dynamics simulations, 60, 70, 76, 77, 80
Multicopper oxidases, 23
Multidomain chimeric enzymes, 54
Multidomain proteins, 58, 59
Multifunctional biocatalysts, 3, 5, 87, 93, 103
Multifunctional chimeras, 63, 69, 86, 91, 102, 103

Multifunctional constructs, 63
Multifunctional protein, 4, 51, 53, 68, 70
Multifunctional xylanases, 84
Mutagenesis, 61, 68, 71, 73, 77, 80, 83, 102
Mutations, 61, 65, 73–75, 77–79, 83, 85, 86

N
Nanocellulose, 3, 17, 20, 106, 108
Nanocellulose production, 20
Nanofibrillation, 26–28, 36
Native lipase, 91
Natural enzymes, 1, 3, 7, 25, 40, 68, 105
Nested insertion, 51, 53
NHA, 23
Nitrile hydratase, 56
Nonhydrolytic proteins, 25
Non-wood fibers, 14
Nordic Swan, 16
N-terminal substitution, 60

O
Office paper, 14, 17
Oil palm EFB pulp, 35
Oil palm empty fruit bunches, 23, 27, 31
Omics technologies, 107
Opacity, 5, 17
ORBIT, 61
Oxidoreductase Fusions, 74
Oxidoreductases, 19, 21, 36, 38, 71, 74
Ozone, 15, 40

P
Packaging materials, 1
Packaging paper, 11
*Paenibacillus*sp, 62, 63
Paper, 1–7, 11–15, 17, 19, 21, 22, 25–28, 36–38, 40, 41, 49, 53, 59, 87, 89–95, 100–108
Paperboard, 2
Paper brightness, 22, 23
Pectate lyases, 22, 24
Pectinase, 19, 21, 22, 24, 25, 27, 30, 31, 34, 36–40
Pectin degradation, 22
Pectinolytic enzymes, 28, 31, 93
PELE, 61
Peracetic acid, 15
Persistent pollutants, 4, 15
PFI milling, 40
Phanerochaete chrysosporium, 29
Pharmaceuticals, 4, 53, 64, 68, 69, 71, 74

Phenol oxidases, 21
Phospholipase, 91
Pichia pastoris, 25, 71, 77, 89, 91, 96–98, 100, 101
Pine, 14, 24, 27, 28, 39
Pine kraft, 27
Pitch, 1, 5, 7, 20, 37–39, 88, 93, 103, 105
Pitch control, 1, 5, 7, 20, 37–39, 88, 93, 103, 105
Plant cell, 24, 25, 51, 54, 90
Platform chemicals, 6, 100
Poly(Dimethyl Diallyl Ammonium Chloride) (PDADMAC), 28, 40
Polygalacturonase, 24
Polygalacturonic acid, 22, 24
Polyketide synthases, 82
Polymerase Chain Reaction (PCR), 58, 96–98
Polyoxyethylene ether, 28
Polypeptides, 51, 54, 64, 66, 69, 81, 88, 89, 91, 93, 94, 100, 101
Post-consumer waste, 14
Post-translational conjugation, 66, 96, 99
Predictive model, 71, 74, 79, 81
Printability, 14, 15, 106
Printing and writing paper, 11–13
Programme for the Endorsement of Forest Certification (PEFC), 16
Proline, 50, 54, 70, 73, 80
Prostaglandins, 50, 54
Protein engineering, 1, 6, 49, 51, 53, 54, 59, 68, 69, 75, 94
Proteolytic degradation, 85, 89, 101, 102, 104
Proteomics, 107
Protparam, 62
Psychrophilic, 75
Pulp brightness, 4, 27, 40, 87, 90, 100, 101, 108
Pulping, 1, 4, 6, 7, 11, 13–15, 23, 24, 26, 37, 40, 49, 89, 103
Pulp whiteness, 23
Pulp yellowness, 23
Pyranose oxidase, 23
Pyrroline-5-carboxylate synthase, 50, 54

R
Rational design, 4, 40, 51, 67, 68, 72–74, 77–81, 83, 87, 94, 102
Rational Linker Engineering, 73
Reactivity, 25, 28, 35, 39
Recombinant chimeras, 51

Recombinant DNA, 51, 54
Recombinant techniques, 23, 54
Recyclability, 3, 16, 17, 19
Recycled fiber, 12–14, 108
Recycled paper, 3, 5, 21, 28, 39, 91, 106
Recycled paper processing, 3
Recycling, 1, 5, 12, 28, 41, 91, 106, 108
Refining, 3, 14, 15, 19, 22–25, 27, 28, 34–40, 105, 106, 108
Regulatory chimeras, 93
Resin degradation, 7
Restriction endonuclease, 50
*Rhodococcus*sp, 25
Rice straw pulp, 34
Robust cellulases, 74
Root Mean Square Deviation (RMSD), 79
Rosetta, 61, 73, 77, 78, 80
Rosettadock, 79
Rosetta-engineered β-1,4-xylanase, 61
Rosetta enzyme, 61
Rossmann folds, 75

S
SABER, 61
Scaffold, 61, 72, 75, 76, 79, 102
Scaffold selection, 75, 76, 79
Screening, 14, 51, 71, 73, 77, 78, 80, 82, 83, 86
Secondary fiber processing, 5
Semi-chemical pulping, 14
Serine-histidine-aspartate triad, 74
Sheet strength, 15, 91, 98
Short rigid linkers, 82
SignalP program, 62
Simulation, 16, 60, 70, 73, 74, 76–80
Site-directed mutagenesis, 71, 73
Sludge, 15
Smoothness, 5, 14, 15
SnapGene software, 62
Sodium hydroxide, 14, 34, 87
Sodium silicate, 34
Sodium sulfide, 14
Softwood kraft pulp, 34, 101
Softwoods, 5, 6, 14, 30, 34, 39, 101
Solid waste, 2, 15
Solid waste generation, 2
Specialty papers, 11, 12, 15, 17, 39
Spruce, 14
Stability prediction algorithms, 77
Starch, 13, 22, 23, 36–39, 55
Starch hydrolysis, 22
Sticky control, 3, 19, 22, 24–26, 36

Sticky removal, 7
Straw, 14, 23, 24, 27, 28, 31, 34, 90, 95, 98
Streptomycesthermoviolaceus, 60, 63
Substrate affinity, 6, 63, 66, 69, 88, 92, 100
Substrate specificity, 3, 5, 7, 40, 63, 66–69,
    71, 75, 80, 94, 97, 103, 104
Sulfite process, 14
Surface charge engineering, 77
Surface residue optimization, 76, 77
Surface sizing, 15, 37–39
Surface smoothness, 5
Surfactant, 25, 28, 34, 40, 54, 58, 59, 91
Suspended solids, 15
Sustainability, 2, 3, 5–7, 11, 13, 14, 16, 17,
    20, 21, 28, 101, 103–105, 108
Sustainable certifications, 16
Sustainable development goals, 17
Swelling, 27
Swollenin, 22, 26
Synthetic biology, 1, 6, 40, 71, 74, 81, 87,
    107, 108
Synthetic polymers, 15

T
Tandem fusion, 66, 92
Tensile strength, 30
Therapeutics, 4
Thermal stability, 3, 7, 54, 61, 67, 69, 80,
    95–97
Thermomonospora, 24, 55
Thermophilic catalytic domain, 77
Thermostable, 59–63, 92, 101, 102
Thermostable chimeras, 63
Thermotoga maritima, 55, 60, 63
Three-domain insertions, 51, 53
Tissue, 1, 11, 12, 17, 39
Tissue and hygiene products, 11, 12
Totally Chlorine-Free (TCF), 15, 39
Toxic effluents, 4, 105
Transcriptional regulator C protein, 50
Trichoderma, 24
Trifunctional chimeras, 84
Tumor-targeting antibodies, 64
Two-domain insertions, 51, 53

U
Ultrasonic treatment, 40
Unbleached Kraft pulp, 30

V
Value-added products, 6
Versatile peroxidase, 21, 23
VOC emissions, 24

W
Wastepaper deinking, 25
Water contamination, 2
Wheat straw pulp, 27, 28, 31
Whispering-gallery mode sensors, 4

X
X-ray crystallography, 71, 73
Xylan, 22, 29, 37, 51, 60, 63, 90, 99, 100
Xylanase, 4, 5, 7, 15, 19, 22–31, 33, 34,
    36–41, 53–55, 59–63, 66, 67, 70, 84,
    87, 90–97, 100, 103, 105
Xylanase-acetylxylan esterase, 98
Xylanase-arabinofuranosidase, 96
Xylanase-β-glucosidase, 95
Xylanase–CBD3a fusion, 63
Xylanase–cellulase, 94, 98
Xylanase-feruloyl esterase, 95
Xylanase–laccase hybrids, 103
Xylanase–lichenase, 97
Xylanase-mannanase, 97
Xylanase–mannanase chimera, 63
Xylanase-xylosidase, 97
Xylan degradation, 22
Xylanolytic gene expression, 93
Xylano-pectinolytic enzymes, 31
Xylosidase, 22, 24, 55, 62, 63, 66, 90, 92
Xylosidase/arabinofuranosidase chimera,
    29, 66, 92
XynB-STX-II, 63
XYN-TmCBM9-1_2, 60, 63

Y
Yarrowia lipolytica, 26